COPING WITH COMPUTERS IN THE COCKPIT

Coping with Computers in the Cockpit

Edited by
SIDNEY DEKKER AND ERIK HOLLNAGEL
Linköping University, Sweden

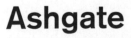

Ashgate

Aldershot • Brookfield USA • Singapore • Sydney

Published by
Ashgate Publishing Ltd
Gower House
Croft Road
Aldershot
Hants GU11 3HR
England

Ashgate Publishing Company
Old Post Road
Brookfield
Vermont 05036
USA

Ashgate website: http://www.ashgate.com

British Library Cataloguing in Publication Data
Coping with computers in the cockpit
    1. Aeronautics - Human factors 2. Airplanes - Automatic
    control
    I. Dekker, Sidney II. Hollnagel, Erik
    629.1'352

Library of Congress Catalog Card Number: 99-65450

ISBN 0 7546 1147 7

Printed and bound in Great Britain by MPG Books Ltd, Bodmin, Cornwall

Contents

BEING THERE: AUTOMATION AND INTERACTION DESIGN

Explains how and why the age-old question of function allocation is no longer adequate to describe the automation technology of today and tomorrow. Provides new sorts of guidance to human-automation design.

Presents the theoretical and practical challenges of designing feedback to make automation a teamplayer, given the realities of actual practice in aviation and related domains.

Addresses how in two-crew cockpits awareness of automation status and
behavior is in part a function of collaboration and accordingly presents an
agenda for cockpit automation / situation awareness research.

GETTING IT TO WORK: CERTIFICATION OF AUTOMATION

Explains from the manufacturer's standpoint how certification of flight management systems fails to catch both very basic and more subtle human-computer interaction flaws. Recommends changes to current practice.

7 HUMAN FACTORS OF AUTOMATION: THE REGULATOR'S CHALLENGE 109

Hazel Courteney

Lays out what the regulators would like to see in terms of systems design and certification on the basis of human factors criteria. Takes into account changing JAR requirements in the near future.

8 EXTRACTING DATA FROM THE FUTURE - ASSESSMENT AND CERTIFICATION OF ENVISIONED SYSTEMS 131
Sidney Dekker and David Woods

Examines the problem of assessing envisioned systems which are as yet ill-defined but likely to create fundamentally new roles for humans, for example air traffic management by exception.

LEARNING TO LIVE WITH AUTOMATION

9 MODERN FLIGHT TRAINING - MANAGING AUTOMATION OR LEARNING TO FLY? 145
Johan Rignér and Sidney Dekker

Reviews approaches to *ab initio* pilot training on the basis that flying automated aircraft is cockpit resource management (managing both automated and human resources). This blurs the traditional distinction between technical and non-technical skills.

10 INTRODUCING FMS AIRCRAFT INTO AIRLINE OPERATIONS
Tom Chidester

Synthesises from a carrier's perspective the research and operational results of flying automated airliners. Sets forth in detail how airlines can adapt their policies and training programs to prepare crews for automated cockpits.

11 AUTOMATION AND ADVANCED CREW RESOURCE MANAGEMENT

Thomas Seamster

Shows how an airline can re-organize its training of crews for cockpit resource management in automated flight decks.

12 AUTOMATION POLICY OR PHILOSOPHY? MANAGEMENT OF AUTOMATION IN THE OPERATIONAL REALITY

Örjan Goteman

Discusses the organizational and technical issues that surround the creation of an automation policy within an airline. Lays out a prototypical automation policy that strikes the balance between being too general (no specific guidance) and too flight deck specific.

1 Computers in the Cockpit: Practical Problems Cloaked as Progress

SIDNEY DEKKER AND ERIK HOLLNAGEL

Linköping University, Sweden

Introduction

Another book on aviation automation? Well, perhaps this is not a book on aviation automation *per se*. It is a book, rather, on how the entire aviation industry is coping with automation. Or more precisely, on how it is coping with the human consequences of automation, which it has fielded over the last two decades. The aviation domain, and the cockpit in particular, is frequently seen to be on the forefront of technological and human-machine interface developments. And indeed, in some sense, progress in the cockpit has been enormous. But from another angle, innovations presented as progress have brought along a large number of unanticipated practical problems - practical problems that today form the inherited by-products of once-vaunted automation technologies. Practical problems cloaked as progress, in other words.

Not only individual pilots have to learn how to operate automated aircraft. The entire aviation industry is learning how to deal with the profound implications that automation carries for the operation, design, regulation and certification of passenger aircraft. The industry is struggling to find ways to meaningfully educate and train operators for their new and different work in the automated environment. It is reconsidering who to select for these jobs and how. And now that current cockpit designs are firmly in place and its problems better-accounted for, it is regrouping to begin to regulate and certify cockpit equipment on the basis of human factors criteria. This while manufacturers are voicing continued concern over the lack of concrete and specific ideas for better feedback design in the next generation of flightdeck automation.

1

One result of being ahead of the pack is that an industry encounters and, hopefully, solves a host of new problems and thereby generates an experience that can be helpful to others. It is therefore quite ironic that many other industries are in fact (re)-discovering similar automation related problems for themselves as they stumble ahead on technology-driven paths. For example, ship bridges are seeing more and more moded automation technology become responsible for navigation and many other on-board tasks. Standardised design of interfaces or system logic does not appear to exist and formal operator training is neither required nor well-organised. The result is that ships have begun to show the same pattern of human-machine breakdowns and automation surprises that were discovered in aviation years ago (see for example the grounding of the *Royal Majesty*, NTSB, 1996). Hence the need for this book: not only is it relevant to exchange experiences and swap lessons across one industry - aviation - it is also critical to show how one industry has to cope with the consequences of its own automation to industries that are poised to adopt similar systems in their operational environments.

Practical problems galore

To some extent research efforts and operational experience are beginning to pay off. In itself, this book is an outflow of the increasing realisation that automation is a mixed blessing. It reflects operational, educational and regulatory countermeasures that were for example inspired by the 1996 U.S. Federal Aviation Administration report on human-automation interfaces onboard modern airliners (FAA, 1996). Closer to the ground, many organisations that deal with complex automation acknowledge that changes in technology can be a double-edged sword. One defence procurement agency for example, says that they must strike a balance between simpler equipment and highly automated equipment. The reason they cite is that the former imposes greater manpower burdens but the latter can create excessive demands on operator skills and training. Such lessons learned indicate that old myths about automation (for instance that it reduces investments in human expertise) are becoming unstuck.

Nevertheless, almost all sectors of the aviation industry are still struggling in one way or another to adapt to the emerging realities of automation technology - to which this entire book is testimony. The training of pilots from the *ab initio* (zero-hour) level upward, for instance, has come to the fore as a key issue relative to automated flight decks (Nash, 1998; Lehman, 1998). Does requisite time in single piston aircraft of light wing loading have anything to do with becoming a jet transport pilot in a world of near sonic, satellite-guided computer-managed flight at 35,000 feet? These questions emerge during a time when European operators and regulators are attempting to harmonise training and licensing standards across the continent and while North-American operators are gradually losing a major source of pilots (the military), with collegiate aviation programs working to fill the gap (NRC, 1997). Questions about preparing pilots for their new supervisory roles do not stop at the *ab initio* level. The debate about optimal training strategies pervades the airline induction (multi-crew, operational procedures) and type-rating stages as well. A new pilot's first encounter with automation is often delayed to late in his or her training. This means it may fall together with the introduction to multi-crew and jet-transport flying, creating excessive learning demands. Telling pilots later on to be careful and not to fall into certain automation traps (a common ingredient in classroom teaching as well as computer-based training - CBT) does little to prevent them from falling into the traps anyway. The end result is that much of the real and exploratory learning about automation is pushed into line-flying.

Automation also erodes the traditional distinction between technical and non-technical skills. This tradition assumes that interactions with the machine can be separated from crew co-ordination. But in fact almost every automated mishap indicates that the two are fundamentally interrelated. Breakdowns occur at the intersection between crew co-ordination and automation operation. Crew resource management training is often thought to be one answer and is by now mandatory. It is also regulated to include some attention to automation. But all too often CRM is left as a non-technical afterthought on top of a parcel of technical skills that pilots are already supposed to have. Air carriers are coming to realise that such crew resource management training will never attain relevance or operational leverage.

Another issue that affects broad sections of the aviation industry is the certification of flight decks (and specifically flight management systems) on the basis of human factors criteria (Harris, 1997; Courteney, 1998). One question is whether we should certify the process (e.g. judging the extent and quality of human factors integration in the design and development process) or the end-product. Meanwhile, manufacturers are working to reconcile the growing demand for user-friendly or human-centred technologies with the real and serious constraints that operate on their design processes. For example, they need to design one platform for multiple cultures or operating environments. But at the same time they are restricted by economic pressures and other limited resource horizons (see e.g. Schwartz, 1998). Another issue concerns standardisation and the reduction of mode complexity onboard modern flight decks. Not all modes are used by all pilots or carriers. This is due to variations in operations and preferences. Still all these modes are available and can contribute to complexity and surprises for operators in certain situations (Woods & Sarter, 1998). One indication of the disarray in this area is that modes which achieve the same purpose have different names on different flight decks (Billings, 1997).

Air traffic control represents another large area in the aviation industry where new technology and automation are purported to help with a variety of human performance problems and efficiency bottlenecks (e.g. Cooper, 1994). But the development of new air traffic management infrastructures is often based on ill-explored assumptions about human performance. For example, a common thought is that human controllers perform best when left to manage only the exceptional situations that either computers or airspace users themselves cannot handle (RTCA, 1995). This notion directly contradicts earlier findings from supervisory control studies (e.g. the 1976 Hoogovens' experience) where far-away operators were pushed into profound dilemmas of when and how to intervene in an ongoing process.

Technology alone cannot solve the problems that technology created

In all of these fields and areas of the aviation system we are easily fooled. Traditional promises of technology continue to sound luring and seem to offer progress towards yet greater safety and efficiency. For example, enhanced ground proximity warning systems will all but eradicate the controlled flight into terrain accident. We become focused on local technological solutions for system-wide, intricate human-machine problems. It is often very tempting to apply a technological solution that targets only a single contributor in the latest highly complex accident. In fact, it is harder take co-ordinated directions that offer real progress in human-centred or task-centred automation than to go with the technological; the latest box in the cockpit that can putatively solve for once and for all the elusive problems of human reliability.

Many of our endeavours remain fundamentally technology centred. Ironically, even in dealing with the consequences of automation that we have already, we emphasise pushing the technological frontier. We frame the debate of how to cope with computers in the cockpit in the technical language of the day. For example, can we not introduce more PC-based instrument flight training to increase training effectiveness while reducing costs? Should we put Head-Up-Displays on all flight decks to improve pilot awareness in bad weather approaches? How can we effectively teach crew resource management skills through computer-based training tools? With every technical question asked (and putatively answered), a vast new realm of cognitive issues and problems is both created and left unexplored. The result, the final design, may beleaguer and surprise the end-user, the practitioner. In turn the practitioners' befuddlement and surprise will be unexpected and puzzling to us. Why did they not like the state-of-the-art technology we offered them? The circle of miscommunication between developer and user is complete.

One reason for this circle, for this lack of progress, is often seen to lie in the difficulties of technology transfer - that is, the transfer of research findings into usable or applicable ideas for system development and system improvement. This book is one attempt to help bridge this gap. It provides yet another forum that brings together industry and scientific research.

Investing in human expertise and automation development

The book echoes two intertwined themes. The first theme explores how and where we should invest in human expertise in order to cope with computers in the cockpit today and tomorrow. It examines how practitioners can deal with the current generation of automated systems, given that these are likely to stay in cockpits for decades to come. It examines how to prepare practitioners for their fundamentally new work of resource management, supervision, delegation and monitoring. For example, various chapters converge on what forms cockpit resource management training could take in an era where flying has become virtually equated with cockpit resource management (managing both human and automated resources to carry out a flight). There are chapters that target more specific phases in a pilot's training career, for instance the *ab initio* phase and the transition training phase. Yet another chapter makes recommendations on how an air carrier can proceduralise the use of automation in terms of how different levels of automation affect crewmember duties, without getting bogged down in details that are too prescriptive or too fleet-specific.

The second theme explores what investments we must make in the development of automated systems. The industry would like to steer the development of additional cockpit equipment and air traffic management systems in human-centred directions - but how? Current processes of development and manufacturing sometimes seem to fail to check for even the most basic human-computer interaction flaws in - for example - flight management systems coming off the line today. Two chapters examine whether and how certification and increased regulation could help by setting up standards to certify new or additional cockpit equipment on the basis of human factors criteria. Although these chapters represent current European state of the art in this respect, much work needs to be done and much more agreement needs to be reached, for example on whether to pursue quantitative measures of human error and human performance in system assessment. Another chapter lays out how we could finally break away from the traditional but unhelpful, dichotomous notion of function allocation in our debates about automation. As this automation is becoming more and more powerful, allocation of *a priori* decomposed functions misses the point. Also, such increasingly powerful automation needs to show feedback about its behaviour, not just its state or currently active mode - an issue targeted in a chapter on automation visualisation. Finally, one chapter looks into the problem of extracting empirical data about the future. As technology development goes ahead, in aviation and in many other fields of human endeavour, it becomes ever more important to be able to evaluate the human factors consequences of novel technological solutions before huge resources are committed to a particular system design. This chapter explains how to generate empirical data relating to human performance in systems that do not yet exist.

Real progress

As automation has brought along many practical problems under the banner of continued progress, the aviation industry is struggling to cope with the human-machine legacy of two decades of largely technology-driven automation. The lessons learned so far and the lessons still to be learned, carry information not only for aviation, but for a number of industries that are opening their doors to similar problems. Real progress across multiple industries, not the kind that cloaks the sequential introduction of practical problems into different worlds of practice, can only be achieved through acknowledging the similarity in challenges that we have created for ourselves. Hopefully this book offers some leads.

Acknowledgements

The initiative for *Coping with Computers in the Cockpit* was developed in the Swedish Centre for Human Factors in Aviation at the Linköping Institute of Technology with the encouragement and support from *Luftfartsinspektionen*, the Swedish Flight Safety Department. The editors owe particular gratitude to the Director of Flight Safety, Arne Axelsson, and the Flight Safety Department's technical monitors of civil aviation human factors work at Linköping University, Kaj Skärstrand and Bo Johansson.

2 Automation and its Impact on Human Cognition

SIDNEY DEKKER AND DAVID WOODS*

Centre for Human Factors in Aviation, Linköping Institute of Technology, Sweden
* The Ohio State University, USA

Introduction

We introduce new technology because we think it helps people perform better. For example, we expect technology - and especially automation technology - to reduce people's workload, improve situation awareness, and decrease the opportunity for human error. Indeed, we value automation for its impact on human cognition. Through aiding the operator's awareness and decision making, new technology can increase system safety and improve the economy or accuracy of operations.

Looking a little closer, it becomes clear that we express the promises of automation technology almost always in quantitative terms. For example, *less* human workload will result if we replace a portion of the human's work with machine activity. And when we give the human less to do - when we shrink the bandwidth of human interference with system operations - we leave *fewer* opportunities for human error. Indeed, this is the traditional idea: that the replacement of human activity with machine activity has no larger consequences on the overall human-machine ensemble. The only thing that is affected is some kind of outcome measure. This outcome measure may be error count, or workload, or economy, and - indeed - all of them are quantifiable, and all of them somehow get better when we introduce automation technology.

Some of these quantitative effects have been realised, but only in a narrow empirical sense. For example, in highly automated systems there is less workload during certain times. There is also less opportunity - or no more opportunity - to make certain kinds of errors. No one is going to stick the wrong punch card into a flight management system: flight management systems don't work on punch cards (although the equivalent of downloading pre-created flight plans from an airline's operational base into an individual FMS does in fact exist).

But our pre-occupation with the quantifiable is distracting. The real and overall impact of technological innovations is qualitative - not quantitative. Humans and technology are not two separate and interchangeable components of a human-machine system. We cannot simply put in a little more of one and leave out a little more of the other. Instead, changes in one have fundamental consequences for the other. Changes in what we make the machine or the human do also have fundamental consequences for the interaction between them, for how they have to behave in relation to one another. The thought that humans and machines are substitutable or interchangeable without creating any larger impact on the human-machine system other than on some quantitative output measure, has turned out to be a myth. We call this the substitution myth (see also the chapter by Hollnagel in this volume).

ie, the zero-sum myth

Automation technology has had a profound impact on the way people in aviation and other systems do their work. And on what kind of work they do in the first place. Indeed, automation technology has fundamentally changed people's tasks, roles and responsibilities. For example, automation has lifted human responsibilities up into the realm of supervisory control. Here activities like delegation and monitoring are crucial. Human assessments and decisions increasingly have to be about the future. Such new work means new knowledge; new expertise and skill requirements. People who remain at work in automated systems have to know and be good at new and different things. Comments from practitioners across the aviation industry reflect these new realities. "I have never been so busy in my life, and someday this automation is going to bite me", says one pilot. "I diverted my attention from flying the aircraft to attend to the intricacies of reprogramming the computer", says another.

Automation technology has also created novel and unprecedented opportunities for human error and opened doors to new forms of system breakdown. For instance, automated airliners have electronic cocoons of protection wrapped around their failure-prone human pilots. These prevent pilots who are flying manually from going outside the normal flight envelope; from going outside the cocoon specified by the engineers. For example, pilots are not able to stall these aircraft, i.e. go beyond a critical angle of attack. But together - jointly - automation and pilots actually can take an aircraft beyond this angle and outside the envelope. Mishaps such as the crash at Nagoya (1994) show a pilot and autopilot fighting each other for control over the aircraft, which in this case lead to extreme pitch excursions far outside the flight envelope.

Particular patterns, persistent problems

The problems that surround practitioners in their interaction with new technology are more than a series of individual glitches. Of course, we can easily label the incidents to "human error" or attribute the handling difficulties to a "learning curve" which is inevitably associated with the introduction of a new generation of technology, whether aircraft (Benoist, 1998) or something else.

But based in part on a series of investigations of practitioner interaction with highly advanced aviation control environments (Sarter & Woods, 1992; 1994b; 1995; Dekker & Woods, in press), we can conclude that the difficulties that practitioners encounter indicate deeper patterns and phenomena. They all concern human-machine interaction. Aviation is not unique in experiencing these problems. Similar kinds of human-machine breakdowns occur in critical-care medicine (Moll van Charante *et al.*, 1993; Obradovich & Woods, 1996; Cook & Woods, 1996), the nuclear field and railways (Hollnagel, 1997) and the maritime world (NTSB, 1996). Various efforts have compiled these and related research results (Woods *et al.*, 1994; Billings, 1997; Abbott *et al.*, 1996). These works constitute markers in our progressive understanding of the effects that automation has on human performance and cognition. They have also begun to point to strategies that can help us deal with the perceived automation learning curve, or the perceived human error problem.

But problems and false promises persist. In this chapter we will first present two typical automation mishaps. They are from different domains but their anatomy has too much in common to ignore. Their similarity affirms that aviation holds no majority stake in automation problems. It also illustrates how the patterns of human-automation breakdown are of a certain kind - similar from one incident to the next, and from one domain to the next. After describing these incidents we will cover, in turn, what these problems and incidents mean for our investments in human expertise and then how knowledge of them should influence the development of additional or new automated equipment. We will try to map our current knowledge in both areas - setting the stage for the rest of the book which takes us further into what we have learned or still can learn from coping with automation.

Typical mishaps with automated systems

One June 10, 1995, a Panamanian passenger ship named *Royal Majesty* left St. Georges in Bermuda. On board were 1509 passengers and crewmembers who had Boston as destination - 677 miles away, of which more than 500 would be over open ocean. Innovations in technology have led to the use of advanced automated systems on modern maritime vessels. Shortly after departure, the ship's navigator set the ship's autopilot in the navigation (NAV) mode. In this mode, the autopilot automatically corrects for the effects of set and drift caused by the sea, wind and current in order to keep the vessel within a preset distance of its programmed track. Not long after departure, when the *Royal Majesty* dropped off the St. Georges harbour pilot, the navigator compared the position data displayed by the GPS (satellite-based) and the Loran (ground/radio-based) positioning systems. He found that the two sets of data indicated positions within about a mile of each other - the expected accuracy in that part of the world. From there on, the *Royal Majesty* followed its programmed track (336 degrees), as indicated on the automatic radar plotting aid. The navigator plotted hourly fixes on charts of the area using position data from the GPS. Loran was used only as a back-up system, and when checked early on, it revealed positions about 1 mile Southeast of the GPS position - nothing unusual.

About 34 hours after departure, the *Royal Majesty* ran aground near Nantucket Island. A quick check revealed that it was about 17 miles off course and that Nantucket Island was actually rather close by. The accident investigation found that the cable leading from the GPS receiver to its antenna had come loose and that the GPS unit (the sole source of navigation input to the autopilot) had defaulted to dead-reckoning (DR) mode about half an hour after departure. In DR mode, there was no more correction for drift. A northeasterly wind had blown the *Royal Majesty* further and further west.

After the fact, the grounding incident looks mysterious to outsiders who have complete knowledge of the actual state of affairs (Woods *et al.*, 1994). The benefit of hindsight allow reviewers to comment things such as:

♦ "How could they have missed X (the DR mode indication), it was *the* critical piece of information?"
♦ "Why didn't they double-check X and Y (GPS against Loran data), it could have avoided the mishap!"
♦ "Why didn't they understand that X (tripping the cable) would lead to Y (default to DR mode), given the inputs, past instructions and internal logic of the system?"

The wake of the *Royal Majesty's* grounding shows a maritime industry trying to make sense of the consequences of advanced integrated ship bridge systems on human performance. In this new operating environment, for example "the crew's failure to detect the ship's errant navigation for more than 34 hours raises serious concerns about the performance of the watch officers and the master... The watch officer is relegated to passively monitoring the status and performance of the automated systems. As a result of passive monitoring, the crewmembers of the *Royal Majesty* missed numerous opportunities to recognise that the GPS was transmitting in DR mode and that the ship had deviated from its intended track... As the grounding of the *Royal Majesty* shows, shipboard automated systems such as the integrated bridge system and the GPS, can have a profound influence on the watchstander's performance" (NTSB, 1997, pp. 30, 34 and 35).

Almost identical exclamations about human performance followed the 1995 crash of a Boeing 757, near Cali Colombia. After accepting a runway change, the crew programmed the automation to fly the aircraft to a beacon at the end of the new runway. Due to internal logic, however, the flight management computer interpreted the instruction as a different waypoint than the one the crew intended, and it made the autopilot commence a left turn to fly to that waypoint instead. The crew became very busy setting the aircraft up for the new arrival, while the aircraft strayed into the mountains off to the side of the valley in which Cali lies. Almost under total automatic control, it hit a mountain a few minutes after the wrongly interpreted computer instruction. From the position of a retrospective outsider, it is hard to understand how the Cali crew could miss so many critical cues during their gentle but quick progression towards disaster. Indeed, the official investigation blamed various controversial crew decisions for the crash. Aeronautica Civil determines that among the probable causes of this accident were:

♦   The flightcrew's failure to adequately plan and execute the approach to runway 19 at Cali and their inadequate use of automation;
♦   Failure of the flightcrew to discontinue the approach into Cali, despite numerous cues alerting them of the inadvisability of continuing the approach;
♦   Failure of the flightcrew to revert to basic radio navigation at the time when the flight management system-assisted navigation became confusing and demanded an excessive workload in a critical phase of flight (Aeronautica Civil, 1996).

Common reactions to failure

Cali and the *Royal Majesty* have much in common. One thing they share (and share with most other mishaps in high-technology human-machine systems) is the reactions they spawn. Reactions to failure can often indicate the extent of our misunderstandings about human performance in complex worlds, and our underestimation of how new technology has changed their operating environments and the work that has to go on within them.

When we go through some of the reactions ("the crew failed to monitor the vertical flight path...adequately plan and execute the approach...discontinue the approach", etc. or "none of the officers determined that the GPS had switched to DR mode ... their monitoring was deficient ...

they continued to miss opportunities to avoid the grounding"), we get the impression that the respective crews had motivational shortcomings. If only they had tried a little harder, then they would have picked up the data critical to their situation and integrated those in their assessments and decisions - so that the disaster could have been averted. Data critical to resolving the situation was available, so what is puzzling to the retrospective outsider is how these data could possibly have been missed. As the director of safety of an aircraft manufacturer put it: "You can incorporate all the human engineering you want in an aircraft. It's not going to work if the human does not want to read what is presented to him, and verify that he hasn't made an error" (quoted in Woods *et al.*, 1994).

Data availability

The common reactions to failure are predicated on a particular assumption about human performance. This is the assumption of data equi-availability - the idea that if in hindsight data can be shown to have been physically available to practitioners, it should have been obvious or picked up by them.

While it is easy to say that a critical piece of data should have been picked up, the feedback properties of the automated systems that bring these data forth often make it very difficult. In fact, automated systems have made it really hard for practitioners to pick up subtle changes in mode or status. According to one pilot, "unless you stare at it, changes can creep in". Ship bridge systems are generally no better. The GPS unit aboard the *Royal Majesty* was mounted across a chart table and only a two tiny letters (DR) notified the user of the currently active mode. The default to DR mode was automatic: it required no user concurrence. Future behaviour of the ship as governed by the DR mode (in terms of a projected track) was shown nowhere. In other words, no representation on the ship bridge showed in one picture where the automation was taking the ship, even though the automation was in charge of the ship's heading for a day and a half.

These properties of automation have created an interesting situation, described by Nadine Sarter in her doctoral work (see Sarter, 1994). Automated systems have become *stronger* (they can carry out long sequences of action without the user providing any inputs): they have increased authority and autonomy ("Here, you take the boat from Bermuda to Boston"). But the same system remains *silent* about what or how well it is doing, and it is difficult to direct ("How do I get it out of this mode?"). The paradox is that the strength of automation increases co-ordination demands, while the properties of the automation's feedback and interface (showing a tiny "DR"; not asking for user concurrence; not annunciating the mode change in any other way) make such co-ordination extremely difficult.

The reason why co-ordination demands go up is that today's powerful automation is no longer a subsystem that can easily be switched on or off. Operators no longer treat automation as a separate component in the larger operational system. Instead they approach automation as an animate partner in systems operations. "What's it doing now? Why is it doing this?" The kind of automation that steers a large ship towards Boston, or directs an aircraft across the Pacific, no longer just *is*. It *behaves*. "How did it get into this mode? How do I stop it from doing this?" These questions indicate how automation has become an agent, capable of pursuing its own goals (getting this boat from here to Boston) by using knowledge about itself and about the world.

Introducing a new agent has enormous effects on human cognition. An agent is in some sense a new team member. This team member can do certain things on its own, but has to be informed and supervised at the same time. It must also communicate about its work and progress. This means that automation imposes co-ordination demands: a new team member means more co-ordination - precisely the kind of co-ordination that today's automated systems are rather bad at (given that they are silent and difficult to direct (Sarter, 1994)).

The dissociation between availability and observability

The silent strength of automated systems produces a dissociation between data availability and data observability. There is a large difference between data that can be shown to have been available in hindsight ("DR" was available, how could they have missed it?"), and data that was actually observed, used and integrated by the crew given their ongoing tasks and attention demands. Observability is a technical term. It refers to the cognitive work that users need to do to extract meaning from available data. Observability refers to processes, the cognitive work, involved in extracting useful information. It results from the interplay between a human user knowing when to look for what information at what point in time and a system that structures data to support attention guidance. How much cognitive work does the user need to do to make sense out of "DR"? How much guidance does "DR" provide? It is a large step from this unspecific, tiny annunciation to the understanding that your boat isn't heading for Boston after all. To understand what "DR" means, not only in general but also in this particular situation, can take significant mental resources. If the mode annunciation is seen in the first place, knowledge has to be called from memory (what was this "DR" again?) and translated into the expected behaviour of the ship. The knowledge of where the boat is headed is not in the world, not observable. It is up to the head of the user to figure out or remember what DR means and what it does to the boat. This is in a sense an example of Don Norman's "conspiracy against memory". "DR" might have been available, but technically it was not observable.

The dissociation between data availability and data observability is especially noticeable in highly dynamic situations. In these cases, novel combinations of factors can push incident evolutions beyond the routine. Practitioners themselves have little control over the pace of process activities and evidence about unfolding conditions gets generated over time. Gaps and uncertainties are common and practitioners have to make assessments and decisions on the basis of partial and ambiguous evidence. The Cali sequence of events represents precisely such an event-driven, busy period. Saying that something should have been obvious in this situation reveals our own ignorance of the demands and activities of people in complex, dynamic domains, where they must juggle multiple interleaving tasks and sift through uncertain and changing evidence. The fact that certain data was available somewhere in the world during these times does not mean that it was relevant to the multiple tasks at hand, that it was expected, that it was understandable, or that it was in a location or format that made it compelling to look at in the first place.

The critical test for level of observability is when annunciations help practitioners notice what they did not expect to see. Or when it helps them notice more than what they were specifically looking for. If a display only shows users what they expect to see or ask for, then it is merely making data available. The measure of true support comes when the representation helps users see or find what they were not explicitly looking for. Increasing machines' autonomy, authority and complexity creates the need for increased observability. The automation characteristics require new forms of feedback, emphasising an integrated dynamic picture of the current situation, automation activities, and how these may evolve in the future. Striving for these larger representational goals also helps designers achieve a balance between underinforming people about automation activities and overwhelming them with details about every minor action.

Common patterns of breakdown

Cali and the *Royal Majesty* represent a common pattern of breakdown between humans and automation. The incident signature is that perfectly functioning machines are flown or sailed into the ground. There is nothing mechanically wrong with these systems. But through a series of persistent and deepening misassessments and miscommunications between human operators and the automation, they evolve towards failure. This kind of accident sequence can be called the "going sour" accident (see Cook, Woods & McDonald, 1991). In this progression, a small event (e.g. an uncommanded turn to the left; a tripped GPS cable) triggers a situation from which it is in principle possible to recover. But through a series of commissions and omissions, misassessments and miscommunications, the human-automation ensemble gradually manages the situation into a serious and risky incident or even accident. In effect, the situation is managed into hazard.

Accidents such as Cali can hardly be classified as controlled flight into terrain (CFIT) anymore. Instead, "typed flight into terrain", or "managed flight into terrain" would better reflect the critical human-machine interactions preceding the mishap. The automated system is handling the aircraft, which has relegated the crew to a supervisory and directory role rather than a controlling one. As the *Royal Majesty* shows, aviation does not hold a patent on the going sour sequence. The gradually managed "radar-assisted collisions" of the seventies (see Perrow, 1984 for some excellent examples) have now made room for "programmed groundings" such as the one near Nantucket.

The going sour progression is consistent with research findings on complex system failure. Very small - even trivial - events (such as a GPS antenna cable kicked loose) can start a progression towards breakdown (see Perrow, 1984). Many other factors, individually insufficient, are jointly necessary to push a system closer to the edge. Simultaneously, system defences need to breached (e.g. the erosion of double-checking Loran and GPS position data) to allow full-scale system breakdown (Reason, 1990). Having authority and autonomy over safety-critical tasks, highly automated systems can sponsor in their own ways this pattern of complex system failure. The going sour scenario is an important accident category which represents a significant portion of the residual risks in aviation.

Much of the management toward breakdown has to do with the fact that automation often does not help during busy periods. In fact, it gets in the way. When there is already a lot to do, automation will give the user even more to do. It will ask for inputs, it may spring surprises. This problem occurs because of a fundamental relationship. The greater the trouble in the underlying system or the higher the tempo of operations, the greater the information processing activities required to cope with the trouble or pace of activities. The more unusual the situation, the higher the tempo of operations, the more demands there will be for monitoring, for attentional control, for information and for communication among team members. This includes human-machine communication and co-ordination. The upshot is that the burden of interacting with the automated system tends to be concentrated at the very times when the practitioner can least afford new tasks, new memory demands or attentional diversions.

Clumsy automation is a label coined by Wiener (1989) to describe such poor co-ordination between the human and machine. The benefits of new technology accrue during workload troughs: when there was already virtually nothing to do, technology will give the user even less to do. But the costs or burdens imposed by the technology (the additional tasks, new knowledge, forcing the user to adopt new cognitive strategies, new communication burdens, new attentional demands) occur during periods of peak workload; during fast-paced periods of high criticality. This creates opportunities for human error and paths to system breakdown that did not exist in simpler systems.

The common outcome: surprise

What Cali and the *Royal Majesty* also share is the psychological quality at the conclusion of the sequence of events. The ground proximity warning near Cali comes as a surprise: it occurs amidst cues and an evolving model of the world that suggest to Tafuri's and Williams' that their flight is safely in the middle of the valley instead of among the mountains. The fundamental surprise is captured beautifully by Gary Larson's cartoon where an aeroplane is hurtling along when one pilot suddenly asks the other, "Say....what's a mountain goat doing way up here in a cloud bank?" Through the going sour progression, the mismatch between the actual position and the crews' mindset has grown so large that the ground proximity warning requires (or triggers) a fundamental reorganisation of reality.

The same fundamental surprise puts an end to the belief of the *Royal Majesty's* crew that the automation was neatly steering their ship through the Boston sealanes. The grounding came as a total surprise - impossible given the model of the world the crew had. Their model of the world placed the vessel out in the open ocean the entire time and ambiguous cues from the world were rationalised to fit their model. For example, the ship was expected to pass buoys at the entrance of the Boston sealanes, but the buoys could not be sighted. This difficulty was attributed to sun glare, and uncertain radar returns were instead interpreted as buoys, approximately in the places where they were expected to be. The real buoys, of course, were miles away. Smaller vessels in the vicinity of the *Royal Majesty* during the last few hours before grounding were asked what they were doing so far out on the ocean. Only when the *Royal Majesty* had grounded did the crew notice that the Loran system showed their ship's actual location (as opposed to the GPS), with Nantucket being only a few miles away.

The outcome of these circumstances can be characterised in large part as "automation surprises" (Sarter, Woods, & Billings, 1997). These are situations where crews are surprised by the actions taken (or not taken) by automatic systems; where a divergence has occurred between what users expect the automated system to do and what it in fact is (or has been) doing. Crews generally do not notice this divergence from displays of data about the state or activities of the automation (recall the problem of "DR" observability). Instead, the misassessment is detected (and thus the point of surprise is reached) on the basis of unexpected and sometimes undesirable behaviour (the grounding at Nantucket, and the ground proximity warning at Cali). With detection hinging on actual behaviour, there may not be sufficient time for successful recovery. Recovery was attempted in both the Cali and *Royal Majesty* case: in the latter by steering hard to starboard, in the former by pulling up and adding full power. But in both cases it was too late.

Often recovery is not too late. The going sour progression may have been set in motion, but certain factors make their circumstances relatively innocuous. For example it may be daylight, or weather conditions may be clear (neither of which was the case in Cali). With such factors present the sequence of events can more easily be re-directed away from bad outcome. Such evolutions *towards* failure produce pre-cursor events that can signal where the vulnerability of the new system lies. They are in some sense "dress rehearsals" for the real accident. The sequences of events that lead one time to an incident and another time to an accident are virtually identical - the signature of human-automation breakdown is the same. Additional factors - defences breached, double-checking eroded - push the system over the edge or not.

The crash of a highly automated airliner near Strasbourg in 1992 illustrates this. The sequence of events that lead to the crash contained a mode error: the automation had gone into vertical speed mode while the crew was providing computer instructions as if it was still in flight path angle mode. The resulting flight path was a much steeper descent than what was planned and expected by the crew. The accident had been preceded by carbon-copy sequences of events that had not led to accidents because of more favourable circumstances:

"British Airways had had an incident early in its A320 operation when the aircraft had inadvertently been flown on Rate of Descent when the pilots thought they were flying Flight Path Angle. This resulted in a ground proximity warning and subsequent go-around. ... It then came to light that another operator had two similar incidents on record. ... I subsequently learned that our own Training Captains had developed some ad hoc specific preventative training to avoid just this sort of event." (Quoted in Woods *et al.*, 1994, pp. 124-125).

But sometimes these problems occur in more vulnerable circumstances. With other contributors present and defences down, events can spiral towards disaster. Fortunately, going sour progressions towards full-scale breakdown are relative rare even in very complex systems. The sequence of events is usually blocked by the expertise of practitioners and their operational systems. Also, the problems that can erode human expertise and its contribution to system reliability are significant only when a collection of complicating factors or exceptional circumstances come together.

In all cases however, the potential for surprising events related to automated systems appears to be greatest when automated systems act on their own without immediately preceding directions from the human crew (e.g. default to DR mode) and when feedback about the activities and future behaviour of the automated system is weak. So what must a user do to prevent being surprised by the automation? From a variety of accident and incident reports, it is easy to collate the "should haves". The user must:

a checklist

- ◆ have an accurate model of how the system works;
- ◆ call to mind the portions of this knowledge that are relevant for the current situation;
- ◆ recall past instructions which may have occurred some time ago and may have been provided by someone else;
- ◆ be aware of the current and projected state of various parameters that are inputs to the automation;
- ◆ monitor the activities of the automated system;
- ◆ integrate all of this information and knowledge together to assess the current and future behaviour of the automated system.

What these injunctions from accident and incident accounts reveal is that automated systems look very different from the perspective of a user who is caught up in an evolving context as compared to the view of a retrospective outsider; the view of an analyst taking a birdseye view with knowledge of outcome. The retrospective outsider will see how the system's behaviour was a direct and natural result of previous instructions and current state. The user in context will instead see a system that appears to do surprising things on its own.

Empirical data and incident reviews make it clear that opaque black box systems are not team players. These systems create fertile soil for automation surprises - again especially when they are strong but silent; when they have high autonomy and authority but low observability. Although often implicated, automation complexity alone is not the culprit. Many design ideas which keep suggesting that hiding complexity is a useful tactic do not really address the deeper interaction issues. The reverse of making all design data about a machine agent available to users is inappropriate - and critiques of this to justify black box systems that hide system activity are red herrings. Users need to know how to work the system in varying operational contexts. The real questions are:

- ◆ What knowledge must the user have at what level of abstraction about how the system works?
- ◆ How does one tie this knowledge to differing operational contexts and situations?

♦   What kinds of representations make agent activities visible in the context of other cognitive activities and tasks the supervisor is engaged in?

These questions begin to point us in the two complementary directions represented in this book. One direction concerns our investments in human expertise. What knowledge and skills are required for automated systems and how and when must we prepare practitioners for them? The other direction focuses on design and certification of automated systems for future use. We now turn to brief descriptions of the issues in each of these areas.

The investment in human expertise

(P. 23)

One of the myths about automation and human cognition is that as investment in automation increases, less investment is needed in human expertise. But increased automation creates new knowledge and skill requirements. According to one pilot, "I can't fly anymore, but I can type fifty words a minute now".

and ironically requires 'hanging on' to the earlier skills for use in "unusual" situations

New skill requirements

Highly automated aircraft demand that pilots are good at managing a suite of more and less automated resources. They need to delegate tasks to the automation (but not always) and supervise the automated system in its conduct of the task. Post-accident reviews appear to emphasise how such supervision is a critical skill ("if only the *Royal Majesty's* officers had regularly compared position information, they should not have missed the discrepancy"). It is easy to think that the crew could have tried harder, that they could have looked a little closer, but that in fact automation has made them "complacent". But appealing to complacency to explain perceived lapses in supervision is not helpful. It focuses on motivational issues that are supposedly the human's problem, while the issue is actually seldom motivational. "Complacency" ignores the ways in which a human-machine ensemble and its mutual working relationship changes fundamentally as a result of automation.

concur

recall McGill

Delegating work to the automation can be just as controversial as the monitoring of it. Post-accident reviews indicate that the delegation of work to the automation is a judgement, rather than just a skill. According to one airline training manager, "The most important thing to learn is when to click it off." In the wake of the Cali crash reviewers wondered whether the captain should have delegated a simple lateral navigation task to the computer and why he persisted so long in trying to put waypoints "in the box". Many think that this persistence ultimately helped put the aircraft on a mountainside. "How and when to (dis)-engage what" has become part of the traditional (but elusive) "good airmanship". As the Washington Post of April 21, 1996 reported: "American, stunned by its first jet crash in 16 years, is planning to end its policy of training pilots to 'sort out' in-flight anomalies with their computers instead of turning off the computers and flying the plane manually. Some airlines, led by Delta's 'turn it off' program, already have reversed philosophy and are using the American crash to reinforce their training". Similarly, reviews of the *Royal Majesty* grounding questioned procedures and operator training which lead the crew to rely exclusively on automated navigation. Indeed, such behaviour seems to flaunt the most obvious standards of good *reliance* sailor practice, whose "fundamental seamanship practices caution against exclusive on any one source of position information for navigation" (NTSB, 1997, p. 33).

New knowledge requirements

Not only are new skills and judgements necessary. In automated systems practitioners are required to know more and different things as compared to the non-automated system. According to one airline, "There is more to know - how it works, but especially how to work the system in different situations." Automation has not only profoundly changed the human expertise require- *∴ impt'ce* ment, it has also made the acquisition of such expertise more difficult (see *of simulation* also the chapter by Howard in this volume). The complexity and unspecific feedback of automation can lead pilots into developing oversimplified or erroneous mental models of the tangled web of automated modes and transition logics. According to one captain in interaction with his mode control panel, "Oh... now it goes into *this* mode - that means I can...uh... I can't ...uh... move the throttles by hand, or...I'm not sure exactly".

Training departments are struggling to teach crews how to manage the automated systems as a resource in different flight situations. Pilots may have trouble getting a particular mode or level of automation to work successfully. Or they may persist too long trying to get this mode of automation to carry out their intentions instead of switching to another or a more direct means to accomplish their flight path management goals. For example, an instructor may ask, "Why didn't you turn it off?" and the pilot may reply: "It didn't do what it was supposed to do so I tried to get it to do what I had programmed it to do." New knowledge and skill demands are most relevant in relatively rare situations where different kinds of factors push events beyond the routine - just those circumstances that are most vulnerable to going sour through a progression of misassessments and miscommunications. This increases the need to practice those kinds of situations.

The *Royal Majesty's* grounding raised questions about human expertise investments of the maritime world. The manufacturer of the ship's autopilot had classroom and simulator training available, but the owner of the *Royal*

Majesty did not purchase any of this training (and wasn't required to either). When the vessel was placed in service, the manufacturer did provide an "orientation" during sea trials to the first complement of officers assigned to the ship. However, of the officers on the *Royal Majesty* at the time of the grounding, only the chief officer had been to that orientation. On-the-job training attempts were made to patch up some of the gaps, but these focused on routine operations only, during which the system worked as expected all the time. No one was fully proficient in using the navigation system and certainly not in managing any of its possible malfunctions. No surprise, really, that "the master did not have any better understanding of the automated navigation system and the functioning of the GPS than the watch officers" (NTSB, 1997, p. 33).

Teaching automation recipes

New technology means more training. For operators, the introduction of automated systems results in a great deal of training demands. All of these must be fit into a small and often shrinking training footprint. The case of the *Royal Majesty* shows how organisations attempt to deal with these training pressures. One tactic is to focus transition training on just a basic set of modes, and how these modes work in routine situations. Often this is all an organisation *can* offer in terms of preparation, because instructors themselves are not proficient in or exposed to the system's broader functionality. This is the teaching of recipes. Recipes prepare practitioners for stereotypical ways of handling frequent situations with only a small subset of the automation. Recipes restrict the range of options and modes taught, and they concentrate on the system's input-output relationships rather than on its internal workings. The remainder of a system's operation is left to be learned on the line, which does not systematically expose crews to its functionality and modes.

Recipes can deal with an organisation's teaching constraints. They enable organisations to sign practitioners off with the minimum requirements met, while reconciling demands on financial or other teaching resources. But pedagogically and operationally, recipes are problematic. They work only if the basics provide a coherent foundation that aids learning the more difficult parts. That is, if they equip practitioners with the appropriate skills for coordinating the automation in more difficult circumstances. The operational environment must also encourage, support and check in some way that learning continues beyond minimum requirements. Obviously, this was not the kind of operational environment in which the *Royal Majesty* was sailing for Boston harbour. And generally, recipes create the ironic situation that training focuses on those parts of the automated systems that are the easiest to learn. The more complicated parts, the surprising mode transitions, the unexpected failure modes, these are all for individuals to learn later on their own. And for them to learn on the line, where slack to recover from going sour progressions may not exist.

The limits of teaching the way we have before

Traditional teaching settings and materials have begun to fall short of the learning targets for highly complex automated systems. One co-pilot, out of the airforce and new to the airline, remarked: "The book alone didn't do much for me: I had to sit in the cockpit with the book on my lap and then tweak the knobs, push the buttons, flick the switches. Only then did the automation start to make sense". The captain, a veteran of the airline sitting left of the co-pilot, replied in genuine surprise, "What, you actually read the book?"

The message is not that pilots who do not read the book are *miscréants* who should be retrained or exiled. The message is that the book does not cut the mustard when it comes to teaching highly dynamic and interrelated material. Interestingly the static, still presentations found in books or flight manuals are mimicked by vaunted Computer Based Training (CBT), which often boils down to a virtual - and just as static - copy of the book. In general, we should stop trying to squeeze more yield from a shrinking investment in human expertise which relies on traditional methods. It will not help prevent the kinds of incidents and accidents that we label human error after the fact.

Escaping from this trap is essential. A first step is to recognise the limits of minimum requirements. Instead, organisations that have fielded automated systems should produce a culture oriented towards continuous learning. Initial or transition training should produce an initial proficiency for managing the suite of automated resources that is commonly available during line operations. This training should serve as the platform for mechanisms or scaffolds that support continued growth of expertise. This emphasis on continuous improvement beyond initial proficiency is needed because highly automated systems increase the knowledge requirements and the range of situations that practitioners must master. Pilots typically take more than a year of continuous line operations to begin to feel fully comfortable on an automated flight deck.

Exploratory learning and mental models — beyond the routine
└── defined by 'concocted' extremes

A major teaching goal is for pilots to develop accurate and useful mental models of their cockpit systems that can be applied effectively across a wide range of possible conditions. Reaching this goal depends among other things on part-task or full mission practice in line oriented situations. This practice must take pilots beyond textbook scenarios where standard procedures necessarily apply. It must take them beyond routine or canonical versions of abnormal situations. This is critical if flexible transfer of knowledge and the

1.
'out of box'
trng

mastery of complexity is at stake. One pilot for example commented: "Most actual engine failures do not resemble what they teach you in the simulator". The scenario (or simulator) should not merely be realistic, but actually elicit and probe the behaviour of interest (e.g. problems with mode awareness or automation supervision). This will allow practitioners to learn where vulnerabilities in operating the system lie.

Scenarios that go beyond the textbook will also allow pilots to discover the boundaries of their own comprehension, their own knowledge. Practitioners often think they know more about their automated systems then they really do; their knowledge is ill-calibrated. This occurs in part because they have not been exposed to a full array of operating conditions.

2,
explora-
tory
trng.

Exploratory learning is another strategy that can help practitioners recalibrate their knowledge and help them acquire flexibly applicable knowledge about how the automation works. Pilots who have access to part-task training devices or even a CD-ROM with the flight management system software on it, can explore and learn in a non-punitive self-paced setting - one of the best ways to help achieve mastery of a complex, dynamic system. This strategy is becoming more commonly employed in airlines. Pilots, in general, want to improve their knowledge and skills, something that is evidenced by pilot-created guides to the flight management system that can be found in several training centres.

The broader question becomes how we can expand the opportunities to practice managing automated resources across a wide variety of situations throughout a pilot's career. In many ways the aviation industry is well prepared to adopt this overall approach. Various stages in a pilot's training career can be identified - each of which is reflected in separate chapters in this book. For example, the *ab initio* or collegiate pilot training programs are discussed by Rignér and Dekker, who investigated the knowledge needs for future pilots under newly proposed joint European flight crew licensing legislation. Tom Seamster, also in this volume, considers the organisational implications of closely integrating CRM (Crew Resource Management) and automation training. The type-rating courses and issues to do with line oriented flight training (LOFT) are covered in the chapter by Chidester. Taken together, these chapters discuss the different training phases that offer different but interacting opportunities to help spread the automation teaching load across a practitioner's educational life-span.

?
1
2
3

erosion of skills
in automated here
is not covered have been)
(but should

Investing in automation development

Research has shown that a very important aspect of high reliability human-machine systems is effective error detection. Error detection is improved by better system feedback. This should especially be feedback about the future *principle:* behaviour of the aircraft, or its systems. Increased complexity can be balanced to an extent with improved feedback. Improving feedback is a critical investment area for improving human performance and guarding against going sour scenarios. But where and how to invest in better feedback?

wrong approach The normal reaction is to address each specific need for better feedback individually - one at a time. Today's cockpit is in some sense a result of precisely such reactions to failure. This piecemeal approach generates more displays, more symbolic codings on displays, more sounds, more alarms. More data will be available, but they will not necessarily be observable. They will not necessarily be effective feedback: the conglomerate of which they are part challenges the crew's ability to focus on and digest what is relevant in a particular situation.

better approach Instead, we need to look at the set of problems which all point to the need for improved feedback to devise an integrated solution. For example, situations where an aircraft is near the edge of its envelope or in an extreme out of trim condition (Toulouse, Nagoya) form a subset of feedback problems that arise when automation is working at the ends of its authority or at the edges of the aircraft's envelope. Today the automation doesn't clearly tell the pilot that this is the case. When control passes back to crew (e.g., through autopilot disconnect) the crew is behind the developing situation and the aircraft attitude may be unrecoverable.

New feedback design targets

principle: New feedback and improved communication about the automation's behaviour are needed. For example, the automation should indicate when it is having trouble handling the situation. It should let the human know when it is taking extreme action or moving towards the outer parts of its authority. It must also be clearly indicated if agents (the pilot and the automation, for example) are in competition for control over the aircraft.

but This specifies a performance target. The development or design question is how to make the system smart enough to communicate this. How do you define what are "extreme" regions of authority in a context sensitive way? When is an agent having trouble in performing a function (but not yet failing to perform)? How and when does one effectively communicate moving towards a limit rather than just invoking a threshold crossing alarm? From experience and research we know some constraints on the answers to these questions. Threshold crossing indications (simple alarms) are not smart enough - thresholds are often set too late or too early. We need a more gradual escalation or staged shift in level or kind of feedback. A trim-in-motion indication (e.g., auditory signal) may very well say too much. We want to indicate trouble in performing the function or extreme action to accomplish the function, not simply any action.

We know from experiences in other domains and with similar systems that there are certain errors that can occur in designing feedback. Misdesigned feedback can talk too much, too soon. Or it can be too silent, speaking up too little, too late as automation moves towards authority limits. In order to escape from these traps, industry needs to develop, test and adopt fundamentally new approaches to inform crews about automation activities. The new concepts need to be oriented towards transitions (highlighting events, mode changes), and towards the future (projecting expected behaviour beyond the current status). Further, there is merit in pursuing pattern-based representations of automation behaviour, which pilots can scan at a glance and pick up possible unexpected or abnormal conditions rather than have to read and integrate each individual piece of data to make an overall assessment.

These options still leave other questions related to feedback open. Often such questions are driven by the technological capability of the day (or tomorrow). For example, should the feedback occur visually or through the auditory channel or through multiple indications? Should this be a separate new indication or integrated into existing displays? Should the indication be of very high perceptual salience; in other words, how strongly should the signal capture pilot attention? Working out these design decisions requires developing prototypes and adjusting the indications in terms of perceptual salience along a temporal dimension (when to communicate) and along a strength dimension (how strongly to communicate) based on crew performance. All of this requires thinking about the signal in the context of other possible signals.

Going sour incidents and accidents provide evidence that improved feedback is needed. Despite the conflict with economic pressures, prudence demands that we begin to make progress on what is better feedback to support better error detection and recovery. To do this we need a collaborative process among manufacturers, carriers, regulators, and researchers to prototype, test in context and adopt new innovations to aid awareness and monitoring. We need to move forward on this to ensure that, when the next window of opportunity opens up, we are ready to provide more observable and comprehensible automation.

Making up the balance

Automation technology has a profound influence on human cognition and human work. It creates new human tasks, requires new interaction skills and new attentional strategies (e.g. tracking mode status, system behaviour). Automation technology also requires crews to possess and apply new kinds of knowledge. Through all of this it creates new human roles and responsibilities. Operators now need to manage and delegate tasks to a suite of automated resources and supervise the conduct of automated work. Automation related incidents and accidents are easily attributed to problems in the delegation of work to the automation ("he should have gone to heading select, not LNAV") or to deficient monitoring of how the automation carries out its instructions ("they should have known they were in open descent mode").

Table 1: Apparent benefits of automation versus real effects on operational personnel

Putative benefit	Real complexity
Better results, same system (substitution)	Transforms practice, the roles of people change
Offloads work	Creates new kinds of cognitive work, often at the wrong time
Focuses user attention on the right answer	More threads to track; makes it harder for practitioners to remain aware of and integrate all of the activities and changes around them
Less knowledge	New knowledge and skill demands
Autonomous machine	Team play with people is critical to success
Same feedback support	New levels and types of feedback are needed to support peoples' new roles
Generic flexibility	Explosion of features, options and modes create new demands, types of errors, and paths towards failure
Reduces human error	Both machines and people are fallible; new problems associated with human-machine co-ordination breakdowns

Behind such inevitable reactions to failure, and behind the details of particular incidents and accidents, there are broad patterns of human-machine co-ordination breakdowns. It is easy to point to human error and say that humans will have to adapt to the new reality and watch out a little closer while they're catching up on the learning curve associated with new systems. Alternatively, we can learn from these breakdowns and see them as signals of new kinds of vulnerabilities. From this we can conclude that we need to better guard against the kind of incident where people and the automation seem to mismanage a minor occurrence or non-routine situation into larger trouble - the going sour scenario.

by design
not
true

Moving forward with our investments in human expertise and in automation development is not easy. Any change will exact costs on the parties involved. Since the benefits are at a system level, it is easy for each party to claim that they should not pay the costs, but that some other part of the industry should. This is particularly easy because going sour incidents by definition involve many local contributing factors. Each case looks like a unique combination of events with human error as the dominant common factor. Since the aggregate safety level of the aviation system is very high it is easy to ignore the threat of the going sour scenario and argue that the status quo is sufficient ("you approved this before"; "it was safe enough before"). But then, progress will crawl to a halt. The question for regulators, manufacturers, and operators instead is how to build the collaborative environment that can enable constructive forward movement. One major goal of this book is to highlight and report on specific areas for such progress.

3 From Function Allocation to Function Congruence

ERIK HOLLNAGEL

Graduate School for Human-Machine Interaction
Linköping University, Sweden

Introduction

The frame of reference for the discussion in this chapter is the interaction between humans - such as pilots, operators, controllers and other kinds of users - and complex technological systems that include significant amounts of automation – such as modern aircraft, fossil and nuclear power production, continuous and discrete manufacturing, telecommunication, etc. The discussion specifically considers the issue of how humans and machines can work together to accomplish a given and common goal. This has traditionally been referred to as the problem of function allocation. The very term function allocation, however, implies that it is meaningful to talk about specific functions and further that it is meaningful to consider whether these functions can be assigned or allocated to definite parts of the system – specifically the humans or the machines. The purpose of this chapter is to consider in some detail these assumptions, and to point out which consequences they have.

The verb **allocate** means to distribute something according to a plan or a principle, and comes from the Latin *locáre* (to place) which in turn comes from the noun *locus* or place. The origin of the verb therefore refers to the notion of a place in either a physical or a functional space, i.e., something which has a location or can be associated to a recognisable part of the system. The concept of function allocation refers to the division of labour or the distribution of functions or activities between humans and machines. The starting point is a system that includes both humans and machines, where humans and machines have one or more **goals** in common and therefore have to interact in order for the system to accomplish its functions. A further requirement is that the humans and the machines also have some **functions** in common, i.e., that there are some functions that either humans or machines can accomplish. If that is not the case, then the distribution of functions is given by the very nature of the systems, and there is no need for function allocation as a separate aspect or issue.

As an example, both the pilot and the appropriately named autopilot can maintain the altitude and direction of an aircraft, and it is therefore possible to consider whether one or the other should do it. But when it comes to understanding the spoken communication from the air traffic controllers, only the pilot can do that, and function allocation is therefore not an issue. (This will, of course, all change with the introduction of the datalink, but that is another story.)

Two of the main issues for function allocation are: (1) what humans should do relative to what machines should do, and (2) whether humans or machines are in charge of the situation. The first refers to is the **function allocation** (or function distribution) proper, and the second to the issue of **responsibility**. The two issues are obviously linked, since any allocation of functions implies a distribution of the responsibility as well. The problem is least complicated when function allocation is fixed independently of variations in the system state and working conditions, although it is not easy to solve even then. The responsibility issue quickly becomes complicated when the function allocation is flexible, either because adaptation has been used as a deliberate feature of the system, or – more often - because the system design is incomplete or ambiguous. Although both issues have been most conspicuous in relation to the use of automation and to the issue of "intelligent" systems - such as expert systems, artificially intelligent agents, etc. - the problem of function allocation is fundamental for any kind of human-machine interaction, from display design to adaptive interfaces.

Function allocation and decomposition

A fundamental feature of all function allocation methods is that they impose, or are based on, a principle of decomposition. The decomposition is applied both to the system, which is described as composed of distinct parts with distinct performance characteristics, and to the higher level functions which are seen as composed of tasks. Authors, such as Moray (1997), have pointed out that tasks are concrete activities whereas functions are more abstract. It is therefore natural for humans to think of work in terms of tasks, but more difficult to think in terms of function – and it is therefore easier to allocate tasks than functions.

Starting a pump or extending the flaps, for instance, are concrete and identifiable tasks or activities. The functions to which the tasks refer or belong (e.g., to regulate a flow or to maintain lift) are less tangible, principally because they usually can be accomplished in several different ways, i.e., by means of several different tasks.

The common, but often unspoken, premise for all discussion of function allocation is the **substitution assumption**, i.e., that a human function can be replaced by a machine function. This is taken to be true without further consideration – while the opposite clearly is not the case. (For some reason, the problem of replacing a machine function with a human function rarely seems to be an issue. At most, a human function may be proposed or used as a temporary substitute for a machine function that cannot yet be implemented.) To some extent the technological ingenuity and development evident in many systems today gives credibility to this assumption as a praxis, but since it has serious consequences, both in theory and in practice, it is at the very least necessary to examine it further. There are two main reasons why it must be considered with some doubt.

The human as machine. One reason for doubt is that machines, as technological artefacts, function in an algorithmic manner whereas humans patently do not. An algorithm can be defined as a step-by-step procedure that allows complicated operations to be carried out by means of a precisely determined sequence of simpler operations. Although the term is most often applied to software, it is equally applicable to, e.g., electromechanical or electrological systems. Algorithms are also found in the form of procedures (guidelines, checklists) which describe how a goal can be achieved by carrying out a series of simpler operations (e.g., "second right, third left, turn right at the gas station and my house is the third building on the left"). There is unfortunately plenty of evidence that humans do not function very well as machines, i.e., in an algorithmic manner, and that they deviate from the prescribed performance more often than not. This is the case for following procedures (Vicente et al., 1996); for decision making (Lee, 1971); and for reasoning and problem solving (Wason & Johnson-Laird, 1972). Yet replacing part of human functions by machines means precisely that humans are forced to behave machine-like in order to interact with the machines, and therefore embraces the substitution assumption. This reasoning will be developed further below.

Possibly the best illustration that humans function differently than machines is provided by the way in which functions are allocated among humans. Function allocation is a necessary part of human collaboration, either explicitly as functional roles in an organisation (the church or an army), or implicitly as when a group or team organises itself. The assignment of functions to humans is, however, usually made by describing the purpose or goal of **what** the person is supposed to do, rather than described **how** it should be done. Knowing what one has to do, a human is usually able herself to define, organise, and schedule the activities that will achieve the goal – and to adapt the plans to changing conditions. It is only if a person does not know how to do something, that detailed instructions are given; yet this is always the second step, not the first. In contrast to that, machines cannot be instructed by telling them what they should do in terms of the goals they should achieve, but only by telling them how they should do it.[1]

Reason 2 (handwritten margin note)

Performance constancy. Another reason to doubt the substitution assumptions is that human capabilities vary considerable over time. In contrast to that, the capabilities of a machine are constant – or at worst, slowly degrading. The design of joint human-machine systems is based on assumptions about what users know and are able to do. System designers acknowledge that human capacities for perception, attention, discrimination, memory, planning, decision making, and action are limited and therefore try to make estimates which are as reasonable and realistic as possible, to ensure that the performance requirements to the user fall within known limits. Yet even with the most realistic estimates of limitations in human capacity there is an underlying assumption that the capacity is stable or constant. This can be interpreted to mean either that the capacity is invariant with respect to time or that the capacity is invariant with respect to working conditions (cf. Hollnagel, 1990).

In marked contrast to this we all know that there are situations where we function better and situations where we function worse than usual. Looked at in detail (cf. Figure 1) we also know that some functions vary more than others. While physical and physiological functions are reasonably constant, or at least vary in predictable ways, psychological functions, such as attention or mood, and in particular cognitive functions, such as search strategy and short-term recall, may vary considerably.

There may be several causes for these variations. Some of these have to do with the general mood or condition of the user. Thus daily variations in performance (circadian rhythm) are well-documented and regular occurrences. Similarly, there may be changes due to extraneous factors such as the social climate, the personal situation, ambient conditions and weather, etc. On the whole the causes of performance variation will lead to a general decrease (but sometimes also to an improvement) of performance which may be of considerable duration (in the order of several hours or even days). Another set of causes are characteristic for shorter periods of dysfunction, i.e., in the order of hours. Foremost among these are physiological factors (exhaustion), common stimulants (coffee, alcohol, nicotine), etc. Even shorter periods of dysfunction may occur, for instance, because of a temporary overload of tasks, a delay in some actions which leads to an increased time pressure, sudden changes in the working environment (light, noise, movement) which make it difficult to carry out normal procedures, unexpected equipment failure, etc. Taken together these causes mean that it cannot be assumed straightforwardly that human and machine functions are commensurable or even replaceable.

So, if not substitution, then ... (handwritten margin note)

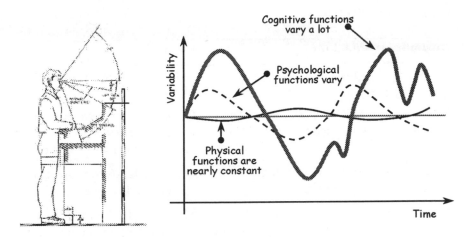

Figure 1: Variations of human capabilities

Substitution versus co-operation

The common basis for function allocation is usually to consider the operator's activities in detail, and evaluate them individually with regard to whether they can be performed better (or more cheaply) by a machine. This approach leads to a decomposition and a consideration of the functions one by one, regardless of the objections to the substitution assumption discussed above. Yet by considering the functions one by one in terms of how they are accomplished rather than in terms of what they achieve, the view of the human-machine system as a whole is invariably lost. The level of decomposition is also arbitrary, since it is defined either by the granularity of the current models of human information processing, or by the machine derived classification from the technological metaphors as epitomised by the list shown in Table 1 (cf. Fitts, 1951). Here the set of attributes are clearly based on machine-like qualities, and do not include more human-like qualities such as being able to filter irrelevant information, scheduling and reallocating activities to meet current constraints, anticipating events, making generalisations and inferences, learning from past experience, collaborating, etc. Many of the human-like qualities are of a heuristic, rather than an algorithmic nature. Admittedly, the use of heuristics sometimes leads to unwanted and unexpected results (also incorrectly referred to as "human errors") but probably no more often than algorithmic procedures do. And the use of heuristics is, after all, the reason why humans are so efficient and why most systems work.

Table 1: Fitts' list

Attribute	Machine	Operator
Speed	Much superior.	Comparatively slow, measured in seconds.
Power output	Much superior in level and consistency.	Comparatively weak, can spurt to about 1500 watts, less than 150 watts over a working day.
Consistency	Ideal for consistent repetitive activity.	Not reliable, always subject to learning and fatigue.
Information capacity	Can be multi-channel. Information transmission in megabits/sec.	Mainly single channel low information transmission rate - usually less than 10 bits/sec.
Memory	Ideal for literal reproduction, access restricted and formal.	Better for principles and strategies, access versatile and innovative.
Reasoning computation	Good deductive, tedious to program. Fast. Accurate. Poor at error correction	Good inductive. Easy to reprogram. Slow. Inaccurate. Good at error correction.
Sensing	Specialised and relatively narrow in range. Good at quantitative assessment. Poor at pattern assessment	Wide energy ranges, some multi-function.
Perceiving	Poor at coping with variation in written and spoken material. Poor at detecting messages in noise.	Good at coping with variation in written and spoken material. Poor at detecting messages in noise.

The determination of which functions should be assigned to humans and which to machines is not as simple as implied by the substitution principle and the categories of Fitts' list. Rather than considering functions one by one, account should also be taken of the nature of the situation, the complexity, and the demands.

"It is commonplace to note the proper apportionment of capabilities in designing the human/computer interface. What humans do best is to integrate disparate data and construct and recognize general patterns. What they do least well is to keep track of large amounts of individual data, particularly when the data are similar in nature and difficult to differentiate. Computers are unsurpassed at responding quickly and predictably under highly structured and constrained rules. But because humans can draw on a vast body of other experience and seemingly unrelated (analogous or "common-sense" reasoning), they are irreplaceable for making decisions in unstructured and uncertain environments." (Rochlin, 1986, p. 4).

Furthermore, the principle of function substitution disregards the higher order need for co-ordination of functions that is required for the joint system to achieve its goals. It maintains the illusions that a substitution of functions is possible despite the long-known facts that humans are notoriously bad as algorithmic processors. There are, of course, some low-level, sensory-motor functions where a substitution is possible and, it might be argued, also desirable, since treating humans as machines is demeaning and should be avoided. In that sense Chapanis was right when he noted that everything that could be mechanised, should be mechanised.

"The nature of systems engineering and the economics of modern life are such that the engineer tries to mechanize or automate every function that can be … This method of attack, counter to what has been written by many human factors specialists, does have a considerable logic behind it … machines can be made to do a great many things faster, more reliably, and with fewer errors than people can … These considerations and the high cost of human labor make it reasonable to mechanize everything that can be mechanized." (Chapanis, 1970).

What Chapanis forgot to point out, however, was that we should only mechanise or automate that which can be completely automated, i.e., where we can guarantee that automation will **always** work correctly and not suddenly require the intervention or support of humans. This can be done only for system states where we are able to anticipate every possible condition and contingency. If automation was confined to those cases, far fewer problems would be encountered. Unfortunately, that is generally not the case.

The arguments listed above lead to the conclusions that function allocation cannot be achieved simply by substituting human functions by technology, nor *vice versa*. Function allocation by substitution is based on a very narrow understanding of the nature of human work and capabilities, which forces a machine-like description to be applied to humans (cf. below). As argued in the preceding, there are really fundamental differences between how humans function and how machines function which makes the substitution principle invalid. A substitution always refers to a very specific view of the function, and disregards the context and the other facets of the situation.

For example, the introduction of alarms is generally considered a step forward. The basic function of an alarm system is to monitor the values of a number of process measurements, and alert the pilot or operator to changes that go above or below predefined thresholds. As such, the alarm system can be seen as taking over or substituting the human task of monitoring the process parameters, and therefore leads to a reduction in the workload or the number of things that the user has to do. While this is undoubtedly true, and alarm systems may certainly be an advantage if they are well designed, transferring the scanning function from the human to the machine also removes the awareness of the process that was a beneficial side-effect. Instead of routinely scanning the instruments on the panel in front of her, the operator can now lean back and relax until the alarm occurs. Yet when it occurs, the first step is to assess the situation, i.e., a new way of responding. Furthermore, the ability to foresee developments and anticipate alarms, may also be reduced or completely be lost. The example therefore shows that one cannot simply transfer the monitoring function from the human to the machine, but must consider the role that the steady monitoring plays for the operators performance and process understanding as a whole.

A specific assignment of functions will invariably have consequences for the whole system. Since functions depend on each other in ways that are more complex than a mechanical decomposition can account for, even apparently small changes will affect the whole - the joint human-machine system. The issue is thus one of overall system design and co-operation between humans and machines, rather than the allocation function as if they were independent entities.

A very short history of function allocation

Historically, function allocation has its roots in the industrial revolution. Before that, artefacts were generally designed for an individual who therefore combined the roles of the designer, the builder, and the user in one person. Although there are some notable exceptions, such as rifles, some tools, pottery, etc., pre-industrial production was on the whole individualistic. In any case tools, such as they were, were designed for discrete rather than continuous processes (rifles, for instance), and their purpose was to accomplish a task directly rather than to control a process or a system performing a task. This all changed as a consequence of the industrial revolution. Firstly, because it became cost efficient to mass-produce artefacts and therefore necessary to consider the characteristics and peculiarities of end users who were not part of the production process. Secondly because the industrial revolution made continuous processes common, it created a need for tools to handle or control them. These developments have, if anything, become more conspicuous with the use of computers and information technology and have created a completely new situation, with radically different and new demands.

The first generation of tools were those that directly affected the process or production, i.e., that directly transported or transformed or manipulated mass, energy or material (or information for that matter) rather than tools that did this indirectly. These tools amplified the human ability physically and directly to manipulate something (in terms of e.g. force, duration, precision, regularity). The new machines could do what hitherto had required human work, and they could do it with greater speed, accuracy or efficiency. Since the speed and complexity of the new tools were limited, it was possible in practice for users to adapt to them, hence to absorb or compensate for any deficiencies in how the tools were designed. Characteristically, little thought was given to the user *qua* tool user, except as a basic philosophy of the role of the human in the production process or the user *qua* worker (e.g., Taylor, 1911; McGregor 1960).

This all changed in the 1950s, due to developments in cybernetics, information theory and computing technology (Wiener, 1954). The scientific and technological innovations made it possible to construct processes and systems that were too complex for unaided humans to handle. It therefore became necessary to consider the inability of the users to meet the increased performance requirements, since this became a limitation of the system as a whole. The new technologies furthered the development of a second generation of tools that affected the processes indirectly and required the users to become controllers that would monitor, control, and regulate the processes rather than directly taking part in them. The problem of function allocation became a critical one, and some important principles were formulated (Fitts, 1951). Looking back from our present stance, the development can be seen as going from design of machine with humans as an appendix and little need to consider human capabilities as a controller (but perhaps as a manipulator), to design of human and machine systems, and finally to the design of joint cognitive systems (Hollnagel & Woods, 1983).

Production, control, and capacity

In the traditional production process the human has to respond quickly enough to meet the demands of the assembly line. But the assembly line is not a process to be controlled. In the production or assembly line the user is directly a part of the material handling and transformation. As such s/he may be a limiting part, but the limitations may easily be overcome. The efficiency of the system, i.e., the output, can be increased simply by having parallel production lines or by dividing the process into ever smaller bits that each can be carried out by more workers. For example, the capacity of a production line can be increased by having shorter activities, such as having two persons each turning a bolt half a turn after each other, instead of having one person turning the bolt a full turn. Having parallel activities can also increase the capacity, since the same result will be produced in the same time – although possibly with an added cost. Common to all of these solutions is the fact that the human only has to understand the single and simple action, rather than how it is part of a whole (i.e., the production process of which the assembly line is a part).

In contrast to this, control of a process cannot be improved simply by having multiple controllers. Even if multiple controllers can be used, they are not independent because they work on the same process. Consider, for instance, air traffic management or process control in general. Any increase in the number of controllers leads to an increase in communication channels and in the volume of information that must be transferred and understood. This sets upper limits on the capacity of the control system. In addition to that, control tasks cannot be made arbitrarily small without losing their meaning. In a sense, the capacity depends on the human ability to comprehend, and this cannot be increased by duplicating the number of humans - at least not without increasing also the amount of communication, hence the time needed to attain and sustain comprehension.

The efficiency of the secondary process (that controls the primary process of manipulation and transformation) can only be improved by having a hierarchy of controllers or of automatic devices. The limit on designing such systems therefore lies in the ability to consider all possible interactions, dependencies, and situations. Since this limit is a very tangible one, the human remains as an essential part of the system. This in turn creates the need to design for human-machine interaction. But if the focus is on the **interaction** between the parts of the system, rather than on the **co-ordination** of the parts of the system (i.e., the control task of the joint system), then function allocation becomes misdirected.

The useless automaton analogy

why milliseconds, as well as multi-seconds, matter

In order to design or analyse the co-ordination of humans and machines it is necessary that either part can be described in sufficient detail and with sufficient precision, and that the two descriptions either use the same semantics or can be translated into a common format or mode of representation. Since the systematic study of human-machine systems started from the need to solve practical, technological problems, the descriptions were based on the engineering view of humans and machines. The technical and engineering fields had developed a powerful vocabulary to describe how machines worked and it was therefore natural to apply this to describe how people worked, i.e., as a basis for modelling human performance (e.g. Stassen, 1986). I will refer to this as the automaton analogy.

why detailed CTA is needed is match to detail the detail of coding

The automaton analogy provides a way to think of or describe a human as an automaton or a machine. A particular case is the use of the information processing metaphor (Newell & Simon, 1972) - or even worse, the assumption that a human **is** an information processing system (as exemplified by Simon, 1972; Newell, 1990). More generally, the automaton analogy has been used in practically all cases where explanations of human performance were sought, e.g. by behaviourism or psychoanalytic theory (cf. Weizenbaum, 1976).

In the field of human-machine interaction, the automaton analogy is, however, useless and even misleading. I will not argue that the automaton analogy is ineffectual as a basis for describing human performance *per se*; I simply take that for granted. (This point of view is certainly not always generally accepted and often not even explicitly stated, for instance, by the mainstream of American Cognitive Science; it is nevertheless a view, which is fairly easy to support.) Instead I will go a step further and argue that the automaton analogy is useless even for machines when the context is human-machine systems, i.e., when the functioning of machines must be seen together with the functioning of humans.

The finite state machine

An automaton is often described as a state-based machine or a finite state machine, meaning that its behaviour can be described by a set of possible states and the corresponding transition conditions. This is closely tied to a description of the allowable input and output for each state. In general, an automaton can be described by a set of inputs, outputs, internal states and the corresponding state transitions; a classical example of this is the Turing machine. More formally, a finite automaton is a quintuple (e.g. Arbib, 1964; Lerner, 1975):

$$A = (I, O, S, \lambda, \delta)$$

Where: I is the set of inputs,
    O is the set of outputs
    S is the set of internal states
    $\lambda: S \times I \rightarrow S$ is the next state function, and
    $\delta: S \times I \rightarrow O$ is the next output function.

Assume, for instance, that the automaton is in a certain state S_j. From that state it can progress to a pre-defined set of other states (S_k, ..., S_m) or produce a pre-defined output (O_j) only if it gets the correct input (I_i). If the system in question is a joint human-machine system, then the user provides some of the input to the automaton. In order for this to work, the user must therefore reply or respond in a way that corresponds to the categories of input that the automaton can interpret. Any other response by the user will lead to one of two cases:

♦ The input may not be understood by the machine, i.e., transition functions λ and δ are not defined for the input. In this case the machine may either do nothing or move to a default state.

♦ The input may be misunderstood by the machine, for instance if the semantics or the syntax of the response have not been rigorously defined. In this case the machine may possibly malfunction, i.e., be forced to a state which has not been anticipated. This happens in particular if the input was a physical control action or manipulation, e.g. like switching something on or off, opening a valve, etc.

The human as automaton

If the machine is to function properly the user must provide a response that falls within a pre-defined set of possible responses (the set of inputs). Yet the determination of the user's response is, at least partly, determined by the available information. The output from the machine therefore constitutes an essential part of the input to the user and that output is partly determined by the previous input i.e., what the user did previously. The content and structure of the machine's output must therefore be such that it can be correctly understood by the user, i.e., be correctly interpreted or mapped onto one of the predefined options for responding (Figure 2). In order to achieve that it is necessary that the designer considers the user as a finite state automaton.

We know from practice that people may interpret information in an almost infinite number of ways. We also know that there is no way in which we can possibly account for this potential infinity of interpretations. Therefore we assume that the user only interprets the information in a limited number of specified ways - and as designers we, of course, take every precaution to prevent misinterpretations from happening. In other words, as designers we try to force the user to function as a finite automaton and we therefore think of the user in terms of a finite automaton. As noted by Rochlin "the more complex and intricate a technical system, the greater the demand it places on operators and technicians to conform inputs to its internal logic" (Rochlin, 1986, p. 5). An excellent example of that is the common graphical user interfaces, which restrict the user's degrees of freedom relative to what a command language allows.

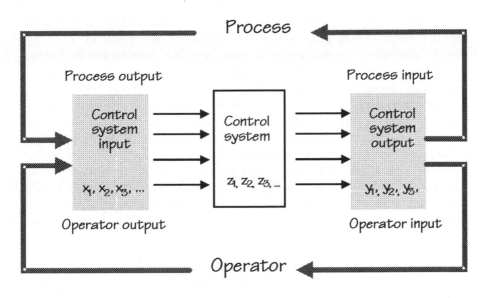

Figure 2: The human as automaton

A more complex example is the use of a machine to buy a train ticket or to cash money. Such transactions are typically carried out by having the user respond to the machine by means of limited sets of choices, presented as menus. The choices furthermore have to be carried out in a certain order or sequence, corresponding to the internal logic of the machine. It is assumed that the users will always correctly understand the output from the machine and be able to find a response that corresponds to their needs. The user is, in fact, treated as a machine with very limited variability. Unfortunately, despite the best efforts of designers, the assumptions often turn out to be wrong, which leads to confusing situations. The process, in this case, is non-dynamic and the worst that can happen is that the user fails to accomplish his or her purpose. Yet it is easy to imagine what could happen if the process was dynamic, and if the consequences were not relatively benign and could affect others than the user. For complex systems we cannot normally expect anyone to comprehend the entire state space, and we cannot protect efficiently against accidents - either incurred by user actions or by component failures.

The starting point of thinking of the machine as a finite automaton has thus forced us willy-nilly to think of the user as a finite automaton, because there is no other conceptualisation or model of the user that will fit the requirements of the design. Yet this is clearly unacceptable, no matter how sophisticated we assume the human automaton to be (even if the number of states is exceedingly large and the state transitions are stochastic or multi-valued).

People, however, are obviously not automata and it appears to be the bane of technological systems that their users do not function as such - despite often courageous attempts to train and drill them to do so. Human action is characterised by a number of things: it is goal directed, it is based on knowledge about the situation, and it can be adapted both locally (operationally, tactically during the action) and globally (strategically, via learning). Since people are not automata, it follows that machines should not be described as automata either. Machines should rather be described similar to humans to enable a proper understanding of the interaction. It is therefore necessary to find a description that captures the essential elements of human action, and use that as a basis for describing both humans and machines. Such a description should avoid the dangers of anthropomorphism as well as of technologising.

Cognitive systems

One proposal for an alternative description is the notion of a cognitive system as defined by Hollnagel & Woods (1983):

> "A cognitive system produces intelligent action, that is, its behaviour is goal oriented, based on symbol manipulation and uses knowledge of the world (heuristic knowledge) for guidance. Furthermore a cognitive system is adaptive and able to view a problem in more than one way. A cognitive system operates using knowledge about itself and the environment in the sense that it is able to plan and modify its actions on the basis of that knowledge." (Hollnagel & Woods, 1983, p. 589).

This definition was based on the experience from working in the field for a number of years, rather than on formal arguments. The definition also reflected the general view of the times, which was to focus on the knowledge processing aspects of humans and machines. A more contemporary definition is that:

> "A cognitive system can modify its pattern of behaviour on the basis of past experience so as to achieve specific anti-entropic ends."

This definition does not go into details of what the components or parts of a cognitive system are, but rather emphasises what the system **does** and which purposes it serves. At the outset this provides a starting point for interpreting the function allocation problem in a way that avoids the problems of the classical interpretation. Although the revised definition is very brief, it entails two important assumptions: firstly that the environment is predictable, and secondly that the cognitive system strives for an equilibrium.

Predictability. Concerning the predictability it is assumed that although the environment is predictable, the system is unable to predict completely how the external world will develop. The reason can be either that the external world includes a stochastic element or is of a stochastic nature. Or it can be that the external world is so complex that the system is unable to overview it in its entirety, hence unable to grasp and handle all the possible relationships. Whether one or the other is the case, the result remains the same. This basically means that the system either must try to survive based on simple reactions, or that it must try to plan its actions and be ready for contingencies. Surviving by prepared reactions is only possible as long as the environment is reasonably stable and the repertoire of (re)actions is adequate. This may be feasible in restricted environments and for a limited period of time, but is insufficient for the kind of complex, dynamic environments considered here. Human performance cannot be based on an algorithmic execution of steps, but must include perception, interpretation, and evaluation - as pointed out already by Miller *et al.*, 1960. In order to survive and be able to achieve its purpose, a cognitive system must have the ability to adapt, i.e., to change behaviour depending on the conditions. The same goes for a joint cognitive system. This means that the system must be able to plan, i.e., to think ahead in time and anticipate the dynamics of the process. It is, however, important to remain on the level of purposeful performance, and avoid a decomposition into the supposedly elementary processes that constitute planning, decision making, and the like.

Equilibrium. The second assumption is that the system will try to restore an equilibrium once it has been disturbed. In fact, rather than always reverting to the same equilibrium the system will try to find an equilibrium that matches the current conditions. Here the important issue is the detection of the discrepancy. This entails that the system must have an internal representation of the desired or intended state, which can be compared to the observed or perceived state. (Note, by the way, that if the discrepancy is not perceived or understood, then the system will be unable to respond appropriately.) Once the discrepancy has been detected and it's nature established, the system will define a goal, corresponding to a new equilibrium. This can also be described as if the system has the intention of doing certain things - i.e., the plans and steps that are assumed to be necessary and sufficient to maintain a homeostasis. The time frame can be long or short and multiple plans can therefore be active at the same time. In order to maintain a homeostasis, the system must continuously act and observe, hence plan and modify the plans. The strategy may not always be known, in which case a trial-and-error approach can be adopted. This will only be successful if the system has the ability to learn, hence to use its experience. Yet we only need to recognise that the system is trying to maintain a homeostasis, and that this provides the driving force, the motivation, and the intentions for its behaviour.

Relative to the issues of designing human-machine systems for process control, cognitive systems engineering provides an alternative to the decomposition approach. Instead of focusing on the components of the joint system and their respective capabilities and limitations, it focuses on the joint system itself. It thereby becomes important how humans and machines co-operate to achieve the joint system function and how they can complement and support each other. Human and machine functions are not seen as being in competition or as being replaceable, but as being mutually dependent and necessary to achieve the overall purpose.

From function allocation to function congruence

As stated in the beginning of this chapter, the concept of function allocation denotes the division of functions or activities between humans and machines, corresponding to the division of labour that was at the heart of the industrial revolution. This definition is usually taken for granted and is therefore an unspoken assumption in all theories or methods. A closer look at the definition, however, reveals that it refers to a particular view on human-machine interaction and therefore embodies a specific set of assumptions about the nature of human performance. Since this view has met with increasing criticism in the 1990s, it is pertinent to consider the consequences of this criticism for the practice and principles of function allocation.

As argued above, the analysis and description of system performance should refer to the goals or functional objectives of the system. The formal definition of a cognitive system was given above. In less technical terms, the definition means that a cognitive system is able to maintain, or re-establish, an equilibrium despite disturbances from the environment. In the context of process control the human-machine systems is seen as a joint cognitive system, which is able to maintain control of the process under a variety of conditions.

The joint cognitive system can obviously be described as being composed of several subsystems, but it is more important to consider how the functions distributed among the various parts of the joint system must correspond and work together in order for overall control to be maintained. Function allocation should therefore be contrasted with a notion of **function congruence** or function matching, which takes into account the dynamics of the situation, specifically the fact that capabilities and needs may vary over time and depend on the situation. One way of accounting for that is to ensure an overlap between functions assigned to the various parts of the system. Each part should be provided or endowed with a set of functions that go beyond the minimum needs, thereby making it possible partly to take over the functions of another part. This corresponds to having a redundancy in the system and provides the ability to redistribute functions according to needs, in a sense to choose from a set of possible function allocations. In order for this to be effective, the functions assigned to various parts of the system must correspond to each other or be **congruous** across the members of the set. The function congruence must be complemented by a set of rules that guides the function re-distribution, keeping in mind the constraints stemming from limited resources and inter-functional dependencies.

An example of function congruence is easily found in human systems, such as the use of human resources in CRM. Examples from technological systems are more difficult to come by, since redundancy usually is avoided rather than encouraged. It may therefore only be found in systems where redundancy is required for safety critical functions or overall mission success (e.g. space exploration). Function congruence in human-machine systems is exemplified by dynamic task allocation, e.g. Vanderhaegen, 1992.

Automation parameters

It is a general experience that any change to the work environment, whether it primarily affects the human, the technological or the organisational factors, will have consequences for the other factors and therefore also for the operators' overall work. In order to be able to establish functional congruence we therefore need a systematic approach to consider the consequences of changes and in particular what the parameters of automation are, comprising: (1) the possible changes to the system and (2) the possible consequences for operator performance.

The question of what are the parameters of automation can be reformulated as a question of what are the parameters of change to the work environment. Such parameters must clearly be described for each separate domain, since they cannot be domain independent. This is partly due to the technology that is used and partly to the nature and constraints of the tasks - the demands of the systems. The parameters of automation (i.e., of possible changes) should furthermore be described in terms of functions rather than structures or components. It is more important to determine **how** changes in the overall system affect performance, than describing **which** changes are made. Since the present concern is about the effects of automation on human work, the parameters must refer to a basic model of human **action** rather than a model of human cognition. The parameters must describe both **which** functions are affected and **how** they are affected. The following concepts provide a useful starting point.

♦ **Amplification**, meaning that the ability to perform a function is being improved. A simple example is the use of machines to amplify physical strength (or speed), e.g., an excavator or digger. Here the person controls every step of the activity, but the physical strength and reach is amplified. In the amplification of cognitive functions the situation is a bit more complicated. Consider, for instance, an expert system for diagnostic support. In some sense it can be seen as an amplification of the person's cognitive functions, such as reasoning, memory, etc. Yet in another sense the amplification easily becomes a substitution, i.e., the system becomes a prosthesis rather than a tool (Reason, 1988). Or rather, if the tool function is not clearly maintained then the complexity of the tool - as well as the fact that it can function partly autonomously - may turn it into a prosthesis. In this case some of the functions become delegated (cf. below) but without control of how they are performed and with little understanding of how the outcomes are produced. In amplification, control is effectively maintained and there is sufficient information to sustain situation understanding.

♦ **Delegation**, meaning that a function is being transferred to another agent or subsystem, but under the control of the user. This effectively means that the user only needs to monitor or control the function, not to perform it. Yet the user is still able to take over at any time, although with a possible cost in terms of reduced efficiency (for instance, inability to meet time constraints). In delegation, control is maintained on a high level, but lost over the details. This follows from the nature of delegation - a task is given to another system, and until a specified goal has been accomplished (a specific sub-task) there is no need or no possibility of controlling the performance.

Within automation, various ways of delegation have been extensively studied (Sheridan, 1992), and Billings (1991) has proposed a classification using the following categories:

- ♦ direct manual control,
- ♦ assisted manual control,
- ♦ shared control,
- ♦ management by delegation,
- ♦ management by consent,
- ♦ management by exception, and
- ♦ autonomous operation.

It is a consequence of delegation that the user does not have detailed control over how the function is carried out. In that sense there is a temporary lack of information or feedback. Control is, however, maintained at a higher level in the sense that the user only needs to be informed when the delegated function has been accomplished - or has failed, as the case may be. The situation understanding does not need to comprise what happens in the sub-system, and lack of information hereof does not have a negative effect.

♦ **Substitution or replacement**. In this case the person in practice not only delegates the function to another agent or subsystem but completely relinquishes control. This easily leads to a loss of situation awareness, which in turn creates an inability to perform the function (a parallel example is that muscles become weaker if they are not regularly used). The effects can be both long term, such as a degradation of process knowledge or competence, and short term such as a reduction in situation understanding. Actually, even amplification leads to some attenuation, but not to the extent that it affects the ability effectively to control the situation. Attenuation is nevertheless the first step towards prosthetic replacement. If the system is well designed the person still retains the overall control, but conditions may easily arise where the ability to take over is jeopardised.

♦ **Extension**, meaning that new functionality is being added. Extension differs from the other modes by introducing a new resource or functionality rather than amplifying or using an existing one. An example of extension is the introduction of a predictive function. Although people to some extent are able to anticipate future events, prediction is not very precise in practice. Introducing a prediction function is therefore not amplification. In principle, it is of course not possible to extend something that does not already exist, i.e., it is difficult, if not outright impossible, to introduce a completely new function. From a pragmatic standpoint it is nevertheless reasonable to say that an extension has taken place if the function was not beforehand exercised in a recognisable way and to a meaningful extent.

To characterise **which** functions are affected, we can use a set of typical cognitive functions such as observation, identification, planning, and action taken either from the Simple Model of Cognition (SMoC; Hollnagel & Cacciabue, 1991) or the Contextual Control Model (COCOM; Hollnagel, 1998). It is clearly possible to use instead a more complex set of functions, although the principles of the approach outlined below should remain the same.

A simple way of combining the **how** with the **which** is to construct a matrix, and characterise each cell. An example of that is shown in Table 2, which also provides an initial attempt of characterising some of the cells.

Table 2: Parameters of automation

	Amplification	Delegation	Substitution	Extension
Observation	Artificial sensors	Automation (closed loop subsystem)	Signal processing, alarm filtering	?
Identification	Signal processing, filtering, logic.	Co-operation, Expert systems (good design)	Expert systems (bad design)	?
Planning	Predictive tools, simulation	Predictive tools, simulation	Guidance	?
Action	Mechanisation	Automatic controllers	Automation, replacement	?

By using these terms it is possible to account for many important aspects of automation. Guidance, for instance, can be described as an example of substitution applied to planning. The purpose of guidance support is always to provide information or specification of the next step to be performed, and the planning and prediction functions are thereby effectively taken over by the system. Properly applied, guidance could be a case of delegation. Improperly applied, it will be a case of substitution.

Using the concepts

In order to assess a specific case of automation, it is necessary to have a clear understanding of what the "nature" of this automation is and how the effects may show themselves. Using the ideas presented above, this can be accomplished in the following ways. (It may help to think of a concrete example, such as computerised procedures.)

The first step is to determine what the automation parameters are, i.e., in what way the automation changes the working conditions of the operator. In the case of computerised procedures, it may be a case of delegation. Basically, some functions are transferred to the computerised procedure system, such as keeping track of the progress, obtaining measurement readings, or even making some of the diagnoses (although this comes dangerously close to being substitution). The intention is clearly that the operator should remain in control of what happens, but that some of the less complicated tasks can be delegated to the computerised procedure system.

The second step is to consider how this may affect the operator's performance. This can be done following the principles of the cyclical performance model, cf. Figure 3, although in practice it may require a detailed cognitive task analysis of the functions being automated. (An example of a cognitive task analysis of procedure usage is found in Hollnagel & Niwa, 1996.) If the computerised procedure works as intended, it should reduce the efforts needed for e.g. observation, keeping track of the procedure, and keeping track of multiple objectives. Altogether this should reduce some of the performance demands.

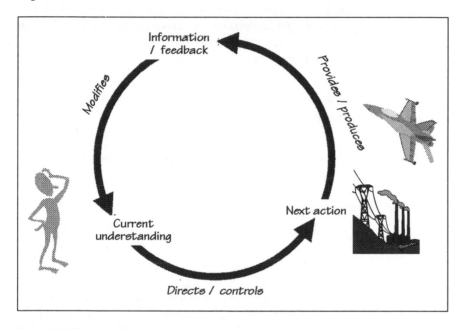

Figure 3: The cyclical performance model

As emphasised by Figure 3, it is essential that the current understanding does not degrade as a result of introducing the computerised procedure system. It must therefore be ensured that the delegation of tasks does not remove information that is essential for the **required** situation understanding. These requirements may, of course, themselves change due to the changes in the task demands.

More concretely, the possible effects on performance can be considered by going through the following set of questions (Table 3):

Table 3: Outline procedure for evaluation of automation

Issue	Explanation
System / function	A specification of the system function that is being changed or modified. This constitutes the initiating factor or the reason for making the analysis. Example: alarm filtering, computerised procedures.
Automation parameter	This refers to the four automation parameters defined above: amplification, delegation, substitution, and extension. The system should be described with regard to which of these are involved or affected, i.e., in which type or level of automation the system provides.
Cognitive functions affected	This refers to the set of cognitive functions that are used in the analysis. These can be either the four simple functions (observation, identification, planning, action) or a different set. They should preferably not bee too detailed.
Intended advantages	This describes the intended advantages of the system, i.e., how it is expected to benefit the work, e.g. by improving specific resources, by changing the performance demands, etc.
Potential disadvantages	This should identify the potential disadvantages that may come from introducing the system. The disadvantages refer not only to that actual work situation, but also to the pre/post-conditions, such as resource needs, long term effects and changes, etc. In general, the intended advantages consider the short term, immediate effects. The potential disadvantages should consider the medium and long term changes that will occur, as the overall system establishes a new equilibrium.

Conclusions: balanced work and automation

A change in the level of automation is often indistinguishable from a change in the part of the technology that is used to perform the task. This can be captured by using the analogy to a balance, although a complex one. The balance is expressed in terms of humans (M), technology (T) and organisation (O) on the one hand, and demands to safety (S) and efficiency (E) on the other. As shown by Figure 4, there is a further balance between (T) and (M) on the one hand, and (O) on the other. As the analogy illustrates, a change in the type of automation (the T-factor) can be functionally equivalent to a change in the M-factor, or a change in the O-factor, relative to the performance demands, (S) and (E). In every case the end result is a disturbance of the overall equilibrium, i.e., a transition to a situation of unbalanced work. If the unbalanced work takes the system outside the region from which it can either re-establish the equilibrium or find a new one, then the ability to perform in a satisfactory manner may have been lost and the situation can be potentially dangerous. Although a change in the T-factor may be functionally equivalent to a change in the M-factor in the sense of the effects on the overall balance, it also introduces a disturbance of the balances between the factors themselves. It will therefore not be sufficient simply to re-establish the overall balance, since there may still be a disturbance of e.g. the (M):(T) equilibrium or the ((M):(T)):(O) equilibrium.

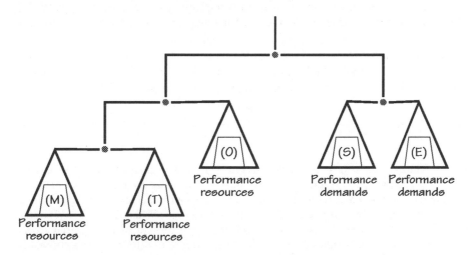

Figure 4: Balance analogy

Although the equilibrium can be disturbed in many ways, changes in the level of automation usually constitute the initiating factor that causes - or rather forces - changes in other parts of the system.[2] Furthermore, the consequences of changing the level of automation are incompletely understood. Quite often people fail to realise that the effects of a change are not confined to the specific function under consideration.

Automation, in the above sense, can involve the pre-processing of measurement data alone and need not involve the automation of actions. Consider, for instance, the case of computerised alarm processing and presentation. (Other examples of the same nature are significant changes to the interface or interaction patterns, to procedures, etc.) A computerised alarm system with alarm filtering is effectively a case of automation in the sense that something that was previously done by the operator now is being done by the computerised alarm system. In a cockpit or control room without alarm filtering all information is presented directly to the operators using a specific technology, e.g., hard-wired annunciators (alarm tiles). This means that all the information in principle is available to the operators, but that it is up to them to scan the alarm information and separate the important alarms from the unimportant ones. This is known to be a considerable burden, and alarm filtering is therefore a frequently used solution.

In a control room with alarm filtering less information is presented because the unimportant alarms have been removed.[3] The general intention is that the operators need not be concerned about **identifying** the important alarms, and that they therefore can spend more time to try to **understand** them and select the appropriate responses. A reduction in the number of alarms is generally considered as an advantage (in particular by the operators), but inevitably has wider consequences for what the operators have to do. The immediate benefit is a reduction in the performance demands because the alarm identification now is done by a technological system, i.e., it has effectively been automated. Although this relieves the operators of a task, it may also deprive them of some information, or have other side-effects such as adding the task of retrieving the suppressed alarms if they are needed. The possible consequences of such side-effects must careful study when a change is planned.

> "The explicit purpose of the new technologies is to increase the number and rate of activities, which already strains human capability to manage, to markedly higher levels. The new system will also be designed to fail less frequently than the old. This has two major implications for operators. When the technical system does fail ..., both the rate and the complexity of the information presented and the difficulty of managing it in order to reassert control over the system will be considerably greater than at present .... Most operators admit that under such circumstances their first strategy will be to shut down as many operations as possible to reduce the overload to tolerable levels. Yet, if the new technology does work as promised to that point, they will have much less practice with coping manually. Thus, technical breakdowns may have far more serious consequences than at present, even if the probability of such breakdowns is greatly reduced." (Rochlin, 1986, p. 6).

The traditional approach to function allocation has been to consider system functions – or tasks – one by one and compare them with the resources or capabilities of identifiable system parts in order to ensure that each function can be carried out in the most efficient manner. In contrast to that, the principle of function congruence emphasises that the functions assigned to various parts of the system must correspond to each other and provide the ability to redistribute functions according to current needs. The balance analogy was a simple illustration of how a change in one part of the system would have consequences for the system as a whole. A proper method of function congruence must be able to describe the dynamic dependencies between functions, and how the overall co-ordination can be ensured, keeping in mind that the primary objective is the ability of the joint system to maintain control. Even though we are not able to do that at the moment, putting the focus on function congruence rather than function allocation forces us to consider the system as a whole, rather than as an aggregation of components.

Acknowledgement

The work described in this chapter has partly been carried out in a project on function allocation in collaboration with Dr. Yuji Niwa of the Institute of Nuclear Safety Systems (INSS), Japan. The support from INSS is gratefully acknowledged.

End notes

[1]   Norbert Wiener once pointed out that the problem with computers was that they did what they were actually told to do, not what we thought we told them to do.

[2]   One counterexample is the decision to reduce the number of people in a team, which may have consequences for the level of automation. Such examples are, however, the exception rather than the rule.

[3]   This, of course, raises the question about how a distinction between different types of alarms can be made, and whether the filtering is reliable and valid.

4 Visualising Automation Behaviour

MARTIN HOWARD

Linköping University, Sweden

Introduction

Flight deck automation in modern jet aircraft has become increasingly sophisticated. The number of systems which air crews have to understand and interact with has steadily increased over recent decades (Billings, 1997). Each of these systems is the result of an engineering process aimed towards producing affordable and reliable systems. There can be little doubt that this has been successful: Modern aircraft and avionics are spectacularly reliable, dependable and robust (Wiener & Curry, 1980).

At the same time, these systems are exceedingly complex; far too complex for any single individual to design and develop. Individual engineers typically have a 3-5 year university degree, many have extensive work experience, yet automation design and development presents such a complex challenge that it requires teams of such engineers to tackle it successfully. The resulting systems are marvels of technology and testaments to the skills of these people.

But the people who fly aircraft, who work on the flight deck and who use these sophisticated systems are not engineers: they are pilots. The training of pilots is quite different from training of engineers and rightly so. We do not expect people who wish to drive a car to have a degree in mechanical engineering, but to be proficient drivers. Still, cars are relatively unsophisticated vehicles and require the human operator to do all the work, even if aided by power steering and anti-lock brakes.

Flight deck automation - however - is another matter. It can maintain a heading, fly an approach, navigate a route, and exhibit a wealth of other complex behaviour, leaving pilots more in a managerial and supervisory role (Dekker & Woods, Chapter 2 of this volume). In order for this to be successful, pilots need to understand what the automation is doing and how it is behaving, if they are to be able to determine when and how to intervene. This should be possible without requiring pilots to become engineers - the level of knowledge needed to operate a piece of equipment should not be equal to that needed to design it. And yet, it almost seems as though one *needs* to be an engineer to understand modern flight deck automation behaviour.

The reasons for this are twofold: (a) it has to do with how automation is designed and developed: I shall examine the philosophical foundations of engineering closer and some of the consequences this has for the artefacts themselves; (2) it also has to do with how we understand and interact with objects in everyday life. How do we make sense of the world around us? By examining these two issues, we may begin to understand why automation sometimes surprises pilots and what should be done to avoid this.

Automation engineering

By 'engineering', we mean the application of general scientific principles to a specific problem, in this case the design and manufacturing of equipment - typically solid state electronics - that performs certain functions. Engineers in turn rely upon physics to model and predict aircraft behaviour, electronics and cybernetics to model and design control circuits, and mathematics and computer science to program some of the components. At each stage, the aim is to - prior to building anything - model and predict the outcome of the system. Engineers work towards quantitatively expressed goals and use empirical methods to test whether a particular design, often expressed in the form of a prototype, achieves these goals or not.

But engineering also means more than this: Engineering problems are often very complex and need to be broken down into constituent parts to be manageable. Each part can then be tackled by a different person or team and the resulting solutions for each part can be synthesised into a solution for the original problem. Given that everything has to follow the laws of physics, we can rest assured that, provided each component solutions works and they fit together, the overall solution will work too. This is obviously a very cost effective approach: We can design different parts in parallel without knowledge of the inner workings of the other, as long as we understand how they are to finally fit together. This enables a top-down approach were detailed design is deferred to the experts in each area, while ensuring that we early on have some grasp of the final cost, performance and capability of the whole design.

In philosophical terms, this is a *rationalistic* approach. It can be summarised as a series of steps:

♦ Characterise the situation in terms of identifiable objects with well-defined properties.

- Find general rules that apply to situations in terms of those objects and properties.
- Apply the rules logically to the situation of concern, drawing conclusions about what should be done. (Winograd & Flores, 1986).

We can also characterise the engineering process with other words: It is *deductive*, because we predict performance of future artefacts based upon models deduced from natural science theories; it is *positivistic*, because we employ experimentation and observation as the only reliable sources of knowing how prototypes and products perform; it is *objective*, because we use repeatable, quantitative measurement to determine performance, rather than assessment or judgement by individuals; it is *analytical*, because the process is decompositional, both with regard to understanding the problem and designing the solutions.

From this we can deduce some properties of engineering problems and products:

- Engineering design and development concerns itself only with problems that can be expressed in terms of (natural) science and with objectively measurable, quantitative design features. [rationalism, positivism]
- Performance specifications are deduced from theoretical models and the specification of the initial problem. [*rationalism, deduction*]
- Only design features that logically follow from principles of (natural) science will be suggested. [*rationalism*]
- Once design features meet the performance specification, they will be refined no further.
- The final product will be a synthesis of solutions to sub-problems to the initial problem. [*analysis, synthesis*]

Point 4 follows not so much from a philosophical characteristic, as the overall aim of engineering:

> "Engineering is not the process of building a perfect system with infinite resources. Rather, engineering is the process of economically building a working system that fulfills a need." (Good *et al.*, 1986, p. 241).

Taken together, these characteristics have two important implications:

- Engineering is artefact-centred: the aim is to design, develop and manufacture products. The situation of use is expressed as a list of functional requirements in the problem specification, but typically not of further concern during the process.

♦ Engineering results in artefacts that primarily make rationalistic sense. They are understood through reason, rather than experience, intuition, or aesthetics.

The artefact-centred nature of this approach means that an analysis of the situation of use is expected to take place *prior* to the engineering design process. Indeed, it provides the input to this process, through identifying the need for a system and expressing this in the form of specifications and requirements, which in turn provide a scope and limitations for the initial analytical stages. The design and development is then concerned with how those specifications and requirements are fulfilled, but not too concerned with what they are: We assume that we are building the right thing, since someone ordered it, but are very concerned with building it right.

Since, working as engineers, we can allow ourselves to focus on artefacts, taking the situation of use as given, and since we use a rationalistic approach, the final products will make 'rationalistic sense'. We can understand how the product works at any given situation by deducing this from the cues it presents and the underlying principles of operation. We expect people to use and interact with the product in a logical, rational way: if certain information is requested, the correct information should be entered. If information is presented on a display, we expect it to be read and understood. When we talk about someone being 'professional', it is often this we mean: That person is acting rationally and logically; they are knowledgeable and competent in their work and well versed the principles and methods that form the basis of that work. Professionals do not act out of whim, they do things for good reason.

But what makes logical, rational sense in one setting may not make sense in another. The comparatively time-consuming, analytical, deductive reasoning in technical rationality - as used by engineers - is not always possible in dynamic, event-driven, process control environments - as experienced by pilots. The situation of design and the situation of use are qualitatively very different and, since the engineering process is not primarily concerned with the situation of use, there is an opportunity for artefacts designed in the one environment to be poorly adapted to the other. We have identified one of the contributing factors to why automation is not always 'used correctly': it has been designed for the wrong type of usage.

Automation use

This is an indication that we need to use a different theoretical model for our analysis and understanding of crew behaviour. Conveniently disregarding the vast differences in the environment in which this analysis takes place, allows us to apply the time-consuming, fine-grained, deductive reasoning of technical rationality. Yet this is not always applicable for the crews using the automation. In many cases, crew training does not provide the detailed knowledge of inner mechanisms of automation, so they do not have access to the underlying principles of automation operation. Pilots have to partly rely on learning this 'on-the-job': Initial training complemented by experience, means pilots learn to recognise a number of common situations and develop recipes for dealing with them (Dekker & Woods, this volume). As situations are encountered, these are compared to a repertoire of known examples gained through experience and suitable actions are taken. This situated use of knowledge is what Donald Schön calls 'knowledge-in-action' (Schön, 1983). If the outcome is the expected, this affirms and reinforces belief in previously learnt knowledge.

However, the outcome is not always the expected: At this point, we start to reflect over the characteristics of the situation, the actions we chose, the outcome, and the relationship between all of these, which Schön denotes 'reflection-in-action' (*ibid.*). Time and events permitting, this reflection-in-action will eventually lead to a more nuanced repertoire, enabling the practitioners to deal with subtle variations in situations and leading to greater success in the choice of actions.

The key lesson is that for many professions, 'normal' work does not largely consist of deductive reasoning: technical rationality is a poor explanatory model for how many typical working situations are handled (*ibid.*). Practitioners rarely extrapolate from the current situation to previously learnt general principles and then apply deductive reasoning to determine suitable actions. For the vast majority of situations, there is neither the time, nor the necessity: It would constitute an enormous overkill, given the resources available and constraints present, if indeed permissible by those same constraints. The problem at hand is not to be able to trace the correct theoretical origins of an encountered situation, but to *handle* it successfully. Practitioners do not solve problems; they manage situations.

An important portion of managing a situation consists of making decisions. In studies of decision making, formal approaches were long favoured. In these, the decision maker has full knowledge of all available alternatives and, after weighing the likely consequences of each, makes their choice to ensure the optimal outcome. This logical, rational model accounts rather poorly for many decisions taken in everyday life, so alternate models that better account for everyday human behaviour have been proposed - these are collectively called naturalistic decision making, NDM (Orasanu & Connolly, 1993).

There is a parallel between knowledge-in-action and the concept of NDM. Orasanu and Connolly (*ibid.*) state that NDM is 'schema-driven', rather than driven by any computational algorithm: Situations are encountered that call for some kind of decision; features of the current situation are compared to previously encountered situations; a suitable course of action is chosen.

Rather than generating several possible courses of action and evaluating the outcome of each, people classify a situation and select a response known to be suitable for that category of problems. Known as the Recognition-Primed Decision (RPD) model, this accounts better for people's reports of how they work when compared to more analytical decision models (Klein, 1993). Cast into different language, both NDM and the RPD model describe how practitioners use knowledge-in-action to make decisions and reflection-in-action over their careers to build a repertoire of known situations and responses. We do not understand the world around us in terms of a disembodied cognition that observes an objective reality: It is through our existence in and interaction with the world - and situations and artefacts in it - that we understand it.[1]

The importance of experience

In NDM, experience plays a key part. This helps practitioners to understand and recognise situations, and to select a strategy for handling it.

> "Experience enables a person to understand a situation in terms of plausible goals, relevant cues, expectancies, and typical actions." (Klein, *op. cit.*, p. 147).

Two important aspects of this is the direction of attention towards 'relevant clues' and the 'expectancies'. Experienced people do not only observe a situation, but know which cues to look for to enable rapid classification of the situation. This is not passive perception; this is active perception. In the words of Ulrich Neisser:

> "In my view, the cognitive structures crucial for vision are the anticipatory schemata that prepare the perceiver to accept certain kinds of information rather than others and thus control the activity of looking. Because we can see only what we know how to look for, it is these schemata (together with the information actually available) that determine what will be perceived." (Neisser, 1976, p. 20).

In response to the internal information processing class of models that account poorly for everyday observed cognition, Neisser presents the 'perceptual cycle' (see Figure 1).

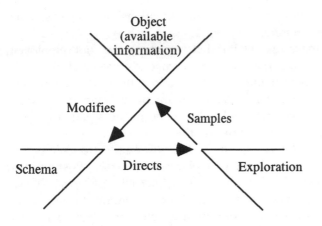

Figure 1: The perceptual cycle (adapted from Neisser, 1976)

The schema, i.e. our previously gained knowledge or experience, helps us to understand the world around us, by directing the exploration of that world and which parts we choose to look at (sample). In return, the world around us may modify our schema, if we observe things that do not fit our current schema. Again, we can see parallels between this and Schön's concepts of knowledge-in-action and reflection-in-action: our learning is directed by what we already know.

Automation use meets automation design

Orasanu and Connolly present eight characteristics of settings where NDM is employed (*op. cit.*, p. 19):

♦ problems are ill-structured;
♦ information is incomplete, ambiguous, or changing;
♦ goals are shifting, ill-defined, or competing;
♦ decisions occur in multiple event-feedback loops;
♦ time constraints exist;
♦ stakes are high;
♦ many participants contribute to the decisions;
♦ the decision makers must balance personal choice with organisational norms and goals.

Many of these also describe work on the modern flight deck. Certainly there are time constraints and stakes are high. There are competing goals -

such as speed, safety, comfort, cost - and both situational and organisational aspects to take into consideration.

Still, how can I make the claim that air crews are not employing analytical, deductive reasoning, but rather knowledge-in-action and NDM? In part, this has to do with crew training. To most pilots, the inner functions of automation is essentially unknown. Instead, patterns of automation behaviour are learnt through experience and on-the-job. Some patterns and responses, i.e. recipes, are pre-learnt: If this happens, do this.

However, the best clues are in the industry's reaction to failure - to mishaps where "human error" is said to play a major role. Post-accident reviews often show that automation was working correctly. By 'correctly', we understand that the equipment was functioning according to its design specification and that it was carrying out the instruction that had been given to it at the time of the accident, without flaws. Yet, viewing the same situation with our new understanding of what work is comprised of, can we say that it functioned 'correctly'? Indeed, can we say it was the 'right' piece of equipment, taking into consideration knowledge-in-action, NDM, and the supervisory and managerial nature of work? If we change our theoretical basis for analysis to one which is more ecologically valid, the results paint a vastly different picture.

Automation exhibits complex behaviour, yet provides few clues as to what it is doing, or how hard it is working. In Neisser's terminology, there are few 'objects' available to pilots to enable them to modify their 'schema' of automation. In other words, many of the seemingly inexplicable features of accidents can be understood in terms of a breakdown in Neisser's perceptual cycle, with repercussions for NDM: Pilots do not know which cues to look for; are too caught up in handling the automation to 'also' fly the aircraft; do not know how to interpret all cues present. Yet a decision needs to be taken and the situation is matched to a previously learnt pattern that provides the closest fit at the time. Sometimes the corresponding course of action fails to manage the situation partly or completely, leading to an incident or accident. In retrospect, the selected action may seem wholly inappropriate, but at the time - to those involved - it makes perfect sense.[2]

From the philosophical perspective outlined above, we can conclude that current automation fails to support the learning and working practices employed by pilots in their new roles as managers and supervisors. The silent nature of automation makes reflection-in-action difficult and dangerous, yet it is through interaction with the world that we come to understand it. Automation does not invite usage. Indeed, the small displays, cryptic command languages, cognitively strenuous interaction, multiple modes, and poor of feedback almost seem designed to discourage human usage.

From a cognitive perspective, current automation does not support perception or decision making. It provides too few clues to enable successful pattern recognition and rapid decision making, or even to build correct schemas of how it functions. It functions as a team member, yet discourages interaction or sharing of information. It is placed in a highly dynamic, event-driven environment, yet implicitly assumes that the human team members will use time-consuming, analytical, deductive reasoning to handle problems.

Consequences for automation design

Unless the goal is to eliminate humans altogether from these environment, we need to approach the problem of automation design from several perspectives. Our focus should not be exclusively on the artefacts themselves, but also on the situation in which they will be used. We need to understand automation not only in terms of build-quality, but also in terms of use-quality.

Ehn and Löwgren introduce the concept *quality-in-use*. It focuses on the actual use situation of an artefact and extends the field of analysis beyond pure constructional aspects to a system under use from three different standpoints: structure, function and form:

> "The *structure* of a system is its material or medial aspects, the technology in terms of hardware and software. The structural aspects of a system are objective in the sense that they are inherent in the construction of the system, and less dependent on context and interpretation.
>
> The *functional* aspects of a system concern its actual, contextual purpose and use. Typically, different users have different purposes for and usage of a system. Organisational dynamics and impacts belong to the functional aspects, as do functions beyond simple utilities of the system: one example is symbolic functions, such as the use of a laptop to signal personal effectiveness.
>
> Finally, the *form* of a system expresses the experience of using the system. Form is not a property of the system, but rather a relation between system and user. This further means that this is subjective, contextual and contingent on the individual user's previous experience." (Ehn & Löwgren, 1997, p. 309).

The authors conclude that to be able to judge a system's quality-of-use is to apply three quality perspectives:

> "The *constructional* quality of the system is expressed in terms of correctness, as is readily exemplified within the field of software engineering. Concepts such as performance, robustness, maintainability and portability are routinely developed and used to talk about the constructional quality of systems, typically in a way that is independent of the current use context.

The *ethical* quality of the system concerns whether it is used in the right way. Ethical questions are typically related to utility and power: Who benefits from the system? Who loses, who wins? Whose purpose is the system fulfilling? Ethical quality is obviously contextual [...]

The *aesthetical* quality of the system is probably the hardest to explain and the least well-developed in the literature. To many individuals, aesthetics is associated with superficial beauty. For instance, a painting is perceived as beautiful, or a dress as pretty. But to take the form aspects of a computer system in use seriously requires more than merely assessing the beauty of the user interface. Again, form is not a property of the system but a relation between system and user. Aesthetic judgement is based on a repertoire of previous experiences ("This word processor feels like a cheap radio"), ideas, values and esthetical concepts such as appropriateness ("the toolbar offers an appropriate set of tools") and comfort." (*ibid.*)

While the authors are primarily addressing information systems development, the concept of quality-in-use is applicable for any artefact with which humans interact. Automation design has been assessed almost exclusively from the perspective of constructional quality. Yet to achieve quality-in-use, we need to consider others too: From an ethical perspective, issues such as responsibility and authority arise. Who is in control? Which team members (both human and machine) have the authority to influence the process? For whom does the automation exist? Pilots or lawyers?

The aesthetical perspective gives rise to questions such as trust, confidence, appropriateness, and suitability. Can I trust that the automation is doing what I want it to? Is it doing the right things? How much effort do I need to put into keeping track of it? Do I feel confident in letting it fly the aircraft?

Visualising automation behaviour

A great deal of time, money and effort has been invested in the latest generation of flight deck automation. One way of improving quality-in-use without resorting to a radical redesign of flight decks, is to try and break the 'strong but silent' signature and improve the communication between humans and machines. Visualising automation behaviour, i.e. trying to show what automation is doing, is one way forward. From the ethical perspective, issues such as authority and control can be made clearer by keeping humans in the loop and ensuring that automation is not being secretive about what it is doing. Focusing on the situation of use and investigating which information is needed, not just from a technical rationality point of view, but from one of knowing and reflecting-in-action, will address automation utility: that automation does the right thing, not just doing it right. From the aesthetic perspective, improved machine-human communication should promote confidence and trust in automation use. Automation surprise and uncertainty are also a important aesthetic qualities which need to be addressed.

Most literature on HCI design is concerned with static display. There is currently little work on real-time visualisation of dynamic behaviour,[3] with a notable exception by David Woods (1994b). Woods calls his approach 'visualising function': How do we construct meaningful representations of dynamic functions, actions, and behaviour, in a computer medium?

When considering existing attempts at providing visualisation for process control, Woods concludes that many fail on a number of key points (*ibid.*). These include:

> *"The keyhole property*, meaning that individual, detailed aspects of the process are accessed through separate screens, with few or no views showing any overview of the process (whole or partial).

> *Context, change and contrast*, key features for discovering process anomalies, are rarely present. 'Disembodied', digital readouts show a current value for some aspect, without reference to 'normal' ranges, or the greater processes they sample.

> *Significance of data* is largely left to operators to discern, through - it is assumed - *the* application of deductive reasoning."

There has been a focus on designing for data display, rather than designing for process visualisation: designs are for *data availability*, rather than for *information extraction* (*ibid.*). Part of remedying this is to separate of the representational form from visual form, according to the symbol mapping principle (emphasis added below):

> "…representational form is defined in terms of how data on the state and *behaviour* of the domain is mapped into the syntax and *dynamics* of visual forms in order to produce information transfer to the agents using the representation, given some task and goal context. The symbol mapping principle means that one cannot understand computer based information displays in terms of purely visual characteristics; the critical properties relate to how data is mapped into the structure and *behaviour* of visual elements. The dynamic aspect of the mapping is critical." (Woods, 1994b, p. C4-1).

Other authors consider the uniquely dynamic nature of information technology artefact too. Löwgren and Stolterman (1998) introduce the concept of an artefact's dynamic gestalt (*dynamisk gestalt*). This is a temporal characteristic that provides perceptual patterns and structure to the artefact, which is perceived and understood over time as a whole. A consequence of this is that all interactive IT-artefacts will have a dynamic gestalt: whether it is coherent, comprehensible and suitable is another matter.

Designing dynamic representations for information extraction is not primarily about designing interfaces: it is about designing interactivity. Issues such as context of use and user-involvement become critical: Central tasks are not being able to read individual pieces of data, but to see aggregates, trends, and relations; to recognise patterns and behaviour, in order to know when and how to act. A move from data-oriented towards process-oriented views, means a shift from communicating numbers and values to communicating issues and ideas (Woods, *op. cit.*). In such displays, the ability to see that parameter x has a current value of 47 is much less interesting than being able to visualise emergent properties, significant patterns, aggregated trends, and integrated wholes. The fact that these properties are hard[4] to express as an engineering specification does not mean that they are uninteresting or insignificant.

Conclusions

We need to consider flight crew actions not in terms of technical rationality, but as knowledge-in-action and naturalistic decision making. We need to design automation to ensure high quality-in-use, not just high robustness and reliability. Our unit of analysis should not be humans or machines, but rather the joint cognitive system they together comprise, the situation of use, and the situated context in which this takes place.

A consequence of this is the need to enrich the communication between these team members. Visualising automation behaviour is one way to improve this communication by providing greater insight into what automation is actually doing. While the exact design solutions are yet unknown, I propose the following list as some of the more important of characteristics of use that these should provide:

♦ ability to show significance, change, trend, and behaviour;
♦ ability to discern emergent features and integrated wholes;
♦ dynamic characters of automation that are in harmony with those of the underlying process;
♦ displays which are relevant for the work performed and understandable in the context of use by those present;
♦ ethical qualities that ensure that authority and control remains with the people *in the context of use* and not just those removed from this context;
♦ aesthetical qualities such as suitability and appropriateness, that promotes confidence and minimises surprise;

◆ a shift from 'strong and silent' to 'understandable and informative' automation.

Success is ultimately going to depend upon our ability to approach these issues as ill-defined problems of interaction design and not well-defined problems of display engineering (Ehn & Löwgren, 1997). The issue is not to replace engineering in automation development: constructional quality is obviously still of great concern, but it is only one of three components that need to be considered holistically. A complementary approach is needed: achieving quality-in-use requires design ability if it is to be successful.

End notes

1   In philosophical terms, this is what Martin Heidegger describes as 'being-in-the-world' (*In-der-Welt-sein*). See, e.g., Steiner (1978) or Cooper (1996).

2   Cf. the concepts of 'local rationality' (Woods *et al.*, 1993) and 'satisficing' (Simon, 1969).

3   Post-test visualization of physical measurements falls outside this definition. It typically does not take place in real-time, and the aim is to analytically understand a sequence of events or establish causal relationships, rather than to act in an event-driven, dynamic environment.

4   But not impossible: See Wenkebach, *et al.* (1992) for an example.

5 Automation and Situation Awareness – Pushing the Research Frontier

SIDNEY DEKKER AND JUDITH ORASANU*

Centre for Human Factors in Aviation, Linköping Institute of Technology, Sweden
*NASA-Ames Research Center, CA, USA

Introduction

Flight deck automation has substantially influenced one of the most enduring recent themes in aviation human factors: situation awareness. Many researchers agree that problems with vigilance, poor feedback and ill-understood system logic all interact to undermine an operator's awareness of what their automated systems are doing and will be doing (Wiener & Curry, 1980; Ephrath & Young, 1981; Kessel & Wickens, 1982; Kantowitz & Sorkin, 1983; Parasuraman, 1987; Norman, 1989; Lee & Moray, 1992; Riley, 1994; Endsley & Kiris, 1995; Sarter & Woods, 1995).

In two-crew glass cockpits, this awareness problem is amplified. Each pilot has separate access to the aircraft's computer systems and can instruct and execute any flight path changes without involving the other human (Segal, 1990). As David Woods put it, each pilot has his or her own private workspace of the CDU (Control Display Unit). Making glass cockpits even more vulnerable to awareness problems is the fact that it is harder to see what the other pilot is doing as compared to older cockpits (Wiener, 1989). Humans need to make up (through explicit co-ordination) for what the machine has taken away from them (the visibility of flight path changes made by other people in the system).

"You shall co-ordinate your computer inputs"

As a result, one of the most recited and re-emphasised procedures in multi-crew cockpits is that pilots must co-ordinate computer inputs with one another before they execute them. Through such a procedure to co-ordinate, human behaviour is forced to adapt around design vulnerabilities (which makes sense, we think, because current cockpit designs will be operational for decades to come). But it is a procedure that is hardly consistently followed. It is a procedural bandaid that doesn't stick in practice. As Wiener observed: "Though some carriers have a procedure that requires the captain (or pilot flying, PF) to approve any changes entered into the CDU before they are executed, this is seldom done" (1993, p. 209).

Complacency has nothing to do with it...

Although many would like to think so, it is not complacency that causes pilots to flaunt the procedure. "Complacency" adopts a model of co-ordination breakdown that is based on a lack of crewmember motivation. The idea is that if pilots really dedicated themselves, they could co-ordinate. But they can't be bothered because it doesn't seem to add much if they do and doesn't seem to be too risky if they don't. The computer usually does the right, expected thing—in other words, the complacency is automation induced.

Appealing to a crew's lack of motivation ignores or even misrepresents how automation has profoundly changed the operating environment and the tasks and roles of the people in the system. The allocation of a particular pilot's attention across a data-rich cockpit is not just (if at all) governed by his or her motivation to look at something. Rather, attention is focused by the expectations and information needs that are in turn driven by a pilot's task goals and evolving mental model of the situation (see Neisser, 1976; Woods, 1994b).

...but other factors might

Research on cockpit co-ordination in automated cockpits has revealed that there are various other reasons that can explain crew's "non-compliance" with the rule to co-ordinate computer inputs. The issue with these explanations is that none of them is able to go much further than re-emphasise the rule that crews should really co-ordinate their inputs, despite or because of situational factors and system features that make it rather difficult. We will say more about this in the remainder of this chapter. For now, what are these explanations?

First, there are a variety of flight management systems that do not even have an execute function, that in fact start executing any flight path change straight away. You can hardly ask human crewmembers to co-ordinate their computer inputs if the computer does not leave them with an opportunity to cross-check inputs before the computer is already underway doing its thing. For the procedure to work in these circumstances, pilots will have to discuss their plans for inputs before one of them begins typing. This may leave gaping holes in the other pilot's understanding of what really is going to be entered into the computer.

Second, on FMSs with or without execute functions, both pilots can be on completely different pages at the same time. For instance, one can be deeply engrossed in computer fuel calculations or typing an ACARS (Aviation Communication And Reporting System) note on the keypad while the other is making changes to the flight plan. Computers have profoundly altered the division of work across a cockpit. Traditional patterns of redundancy and double-checking have eroded while at the same time there is probably always something to do on or with the computer— "In the MD-88 you're always fiddling with something" (Wiener, 1993, p. 199).

The third reason why exhorations to co-ordinate computer inputs don't work most of the time is their redundant nature. They are double-checking procedures as opposed to primary procedures for making something work in the first place. Ironically, this means this kind of double-checking is increasingly likely to be set aside in precisely those circumstances where it is most critical: the high-tempo, event driven situation where crewmembers both have multiple tasks to accomplish in a short time—the same situation with greater potential for input errors. This goes back to the basic nature of aircraft automation: it is "clumsy" in the sense that it gives pilots less to do when they had already little or nothing to do, but it gives them more to do—in fact, it gets in the way—when pilots have a lot to do already (Wiener, 1989). The first tasks that then logically fall by the wayside are the ones that are redundant, the ones that don't add any value to getting the work done, the ones that hold up the execution of the input made and delay the aircraft actually turning or climbing or descending or holding.

Fourth, many flight path changes do not occur in high-tempo event-driven, highly dynamic circumstances, but rather at the ebb of cognitive activity; during the much quieter cruise phase. It is customary during cruise to request and receive direct routings. These are routings where one or more waypoints in between are bypassed in order to aim more directly at a destination. Our own observations, as well as those of many other researchers (see Wiener, 1993) indicate that flight crews in non-dynamic conditions often go solo on the computer, without involving the other pilot in precisely what they are typing. Often the other pilot seems to be just happy with this, and another reason may be that both of them have just heard and confirmed with air traffic control the flight path change on the radio. The intentions are unambiguous and presumably shared across the cockpit, so whoever makes the input doesn't matter and doesn't often seem to abide by the traditional pilot flying/pilot-not-flying distinction either. In fact, Wiener reported on the tendency of crewmembers to help one another with making computer inputs - as in, "here let me do this for you". During quiet phases of flight, "when computer input is required there may be a race to see who gets to make the entry" (1993, p. 210). The clear demarcation between pilot flying and pilot not flying is once again blurred. And in any way, there are few compelling reasons for pilots to both go heads down and check one another's homework if all that has to be done is delete one waypoint by going direct to the one after that - and all of this at a cosy 35,000 feet. This then, for pilots adds to the practical backdrop of making computer entries without double-checking them before execution.

In the final balance, many computer inputs made by one pilot are not cross-checked with the other pilot. This occurs because of perfectly reasonable features of the tasks, tools and goals that humans have at the time. Time pressure, multiple interleaving activities, the backdrop of unambiguous normal practice, and even flight management system design can all conspire against the procedure to co-ordinate what is typed and how the result looks on the map display or other indicators.

But what other than "you shall co-ordinate" can we pass on to airlines on the basis of these explanations? How can we go beyond re-emphasising that crews should co-ordinate their computer inputs despite—or especially because of—circumstances and features that continuously conspire against this? The key lies in our experimental access to the types of situations that sponsor a breakdown in joint crew awareness of what the automation is doing or going to do.

The unit of analysis as barrier

The problem of not being able to go much beyond "you shall co-ordinate" lies in part with the unit of analysis in research on cockpit automation awareness breakdowns (one human - one machine). The unit of analysis in this research has often been one operator (e.g. one pilot), in lone interaction with the automated system. For example, it could be shown that one subject pilot can have difficulty keeping track of automation activities when the machine seems to behave on its own and is relatively silent about what it is doing (Sarter, 1994). The practical challenges that face crews on the line, however, lie at a different level of analysis (see Hancock, 1992; Sarter, 1996). Modern aircraft have two human crewmembers and—for all intents and purposes—one automated crewmember. The two human pilots can be occupied with different tasks, even while each has private access to the automation. The automation in turn gives poor feedback about what it has been told to do (Sarter, Woods & Billings, 1997). This means that the breakdown in crew awareness of what the automation is doing to the aircraft can be an intricate, joint function of the tasks and attentional foci of the various human crewmembers, *combined* with the feedback features of their automated team member. Who is aware (or not aware) of what can for example be driven by the evolving mental model and expectations of one pilot, the computer inputs and attentional activities of the other pilot, and the feedback characteristics of the automation. We could call this a three-way breakdown in awareness—one that derives from the intricate interplay between the two crewmembers *and* the automation.

Limitations in empirical access

Research so far has been limited in systematically investigating the factors (e.g. different tasks, mental models) that lie behind breakdowns in awareness in naturalistic, three-crew (two humans, one machine) settings. It has looked at multi-crew situation awareness in automated systems as a straight aggregate of individual situation awareness (one human + one human - one machine), for example in Segal's (1990) analysis. But this too presumes that the flightcrew is a homogeneous unit in interaction with the automation. And to the extent that they are not homogeneous, they *should* be—through increased co-ordination.

Perhaps the operational community is left to implement and police a blunt-force procedure—"you shall co-ordinate"—that is too coarse to capture the intricacy of many in-cockpit situations that challenge crews' awareness of automation and aircraft activities. The procedure functions as a final behavioural barrier against a problem that has no roots in crew behaviour; a problem that is in fact cognitive, that has to do with intentions and expectations.

In this chapter we lay out the research challenge in gaining better insight into the breakdown in multi-crew awareness. We first review one case where the flightcrew did not co-ordinate a computer input, and then look at that case at a deeper level to discover the intricacy of cognitive factors behind the breakdown. This analysis in some sense specifies the empirical test—can our research explain or replicate a breakdown like this? We then review three dominant strands of cockpit awareness research and assess them in light of the test case. Finally, we propose ways to further calibrate this kind of research with additional data-traces and so get insight into the factors and circumstances behind the three-way breakdown.

Meet the test: confusion over Cali

The crash of a Boeing 757 near Cali, Colombia, represents one instance of the lack to co-ordinate computer inputs. It also represents the intricacy of the three-way breakdown. We will tell two stories here—one stereotypical (the crew didn't co-ordinate because of workload and complacency), and the other more cognitively analytical (multiple tasks, diverging mental models of the developing situation).

The stereotypical story: no co-ordination

Flight 965 was on its way from Miami down to Cali, Colombia when it received a runway change from the Cali air traffic controller. In the cockpit were captain Nick Tafuri and first officer Don Williams. Tafuri had been to Cali before. Williams had never been there. He was pilot-flying. Instead of runway 01 (runway to the north) they were now to take runway 19 (runway to the south—straight ahead of them). Flight 965 was cleared to fly the "Rozo 1 arrival", an approach route that would take them to runway 19. Tafuri consulted the approach chart and found a beacon called Rozo just off the end of the runway. After getting permission from the approach controller he instructed the computer to take the aircraft there, by typing "R", the charted identifier code for Rozo:

> Tafuri: "Can American Airlines, uh, 965, go direct to Rozo and then do the Rozo Arrival sir?"

> ATC: "Affirmative. Take the Rozo 1 and runway 19. The wind is calm."

> Tafuri: "All right, Rozo. The Rozo 1 to 19. Thank you, American 965."

Flight 965 had previously been cleared to fly west of the official airway leading to Cali. From that position it was now headed straight for runway 19, while the Rozo 1 arrival lay to the east. The official reading is that Tafuri wanted to make a shortcut and that the controller didn't really know what he was saying yes to by clearing the aircraft to Rozo (Aeronautica Civil, 1996). Basically, Tafuri wanted to go straight ahead to capture the arrival later on (from D21 onwards) rather than twisting to pick it up all the way from Tulua. Tafuri did not vocalise this intention to Williams.

The problem was that due to database logic, the flight management computer interpreted Tafuri's "R" as "Romeo", a beacon 132 miles away near Bogota, the Colombian capital. As soon as the instruction was executed, the aircraft started turning left towards high terrain between Cali and Bogota. The aircraft crashed minutes later, as the two men were trying to get themselves back to the airport at Cali.

Central to the computer entry is not the fact that there may have been confusion over whether R was Rozo or Romeo, but that Tafuri and Williams did not co-ordinate the entry. Since the instruction was to take the aircraft straight ahead (make no lateral change whatsoever) and had just been explicitly confirmed with air traffic control, there was no perceived need (and no time) to co-ordinate the very simple computer entry with Williams. It is this workload related and complacency induced non-compliance with the procedure that precipitates other events and sends the men towards disaster:

> "Investigators were able to identify a series of errors that initiated with the flightcrew's acceptance of the controller's offer to land on runway 19. The flightcrew [had] expressed concern about possible delays and accepted an offer to expedite their approach into Cali (Runway 19).... One of the AA965 pilots selected a direct course to the Romeo NDB, believing that it was the Rozo NDB, and upon executing the selection in the FMS permitted a turn of the airplane towards Romeo, without having verified that it was the correct selection and without having first obtained approval of the other pilot, contrary to AA's procedures.... The flightcrew had insufficient time to prepare for the approach to runway 19 before beginning the approach.... Consequently several necessary steps were performed improperly or not at all..." (Aeronautica Civil, 1996, pp. 29 and 55).

In the stereotypical account, these events led up to a loss of situation awareness on part of the crew. They lacked awareness of fundamental parameters of the approach to runway 19, lacked awareness of where the aircraft was taking them, and lacked awareness of the terrain surrounding them.

The more complex story: different mindsets and expectations

To the superficial observer, the Cali flightcrew lost situation awareness, but a more careful analysis shows none of that. There is no homogeneous unit that as a whole lost situation awareness. Instead, there are two men with different ideas of where the aircraft was to go: two men with different knowledge, different expectations and tasks, and different attentional foci.

The critical data are that flight 965 was to the right of the airway that formed an official part of the Rozo 1 arrival, and that Williams had never before been to Cali. For Williams, whatever detailed computer input made the aircraft turn was irrelevant. What mattered was that the turn was consistent with his expectations of what they had to do to capture the Rozo 1 arrival from their off-airway position. The manoeuvre was so consistent with expectations in fact, that it did not warrant any comment from him as pilot flying. Underlying Williams confirmed expectation and subsequent silence, however, was the assumption that Tafuri, the other human crewmember, shared the expectation of the turn. In other words, in Williams' assessment, Tafuri had made the aircraft verge left to capture the Rozo 1 arrival. Only when the aircraft had turned too far to reasonably capture the arrival, did Williams begin to comment. This is when Tafuri looked up from his other work and noticed the deviation:

Williams: "Yeah, where are we headed now?"

Tafuri: "17.7, ULQ, uh, I don't know. What's this ULQ? What hap…what happened here?"

For Tafuri, surprise followed: the left turn lay completely outside the pale of his expectations. He was not prepared for it, nor had he been prepared to look for evidence of any roll- or lateral deviation. He had programmed the computer to take the aircraft straight ahead so that he himself was free to concentrate on other pressing tasks to do with the new approach route. The outcome represents the canonical automation surprise (Sarter, Woods & Billings, 1997). The pilot thought he had told the automation one thing, but due to misinterpretation of the keyboard command the automation is actually doing another thing. Gradually the two diverge: Tafuri concentrating on other tasks and in his mind heading for Rozo; the automation taking the aircraft to Romeo.

Cali shows that automation awareness is in part a function of collaboration. Williams flagged the unusual aircraft behaviour and drew Tafuri's attention to it. Cali also shows that the ways in which this collaboration can break down can be cognitively so subtle that it easily slips through the cracks in the behavioural procedure "you shall co-ordinate". Two crewmembers can develop and entertain different models about the world based on the cues they receive, pick out as relevant, and endow with meaning. Automation can interact with these different mental models and propagate them all the way to fully fledged and unrecoverable confusion. Automation creates this effect when one crewmember makes computer inputs that (1) are hard to see by the other crewmember and (2) create aircraft behaviour that is consistent with expectations of that other crewmember, even if it is not at all what the programming crewmember had in mind. The problem is allowed to go out of control precisely because neither crewmember suspects anything— even if for different reasons. The one who made the input has more things to do: he has turned attention away from lateral aircraft behaviour (because he just programmed the computer to take care of that) to the benefit of other pressing tasks. The other crewmember has no reason to raise any questions because observed aircraft behaviour matches what it was supposed to be and surely what the other pilot—who made the input—intended.

The events near Cali represent a larger class of crew-co-ordination problems in automated cockpits. Each human crewmember is equipped with his or her own private access to the third, automated crewmember. The third crewmember, in turn, can carry out long sequences of actions by itself with input (not even directly preceding) from one of the human pilots. There are indications that the kinds of going sour scenarios typical of Cali represent a large residual risk in aviation today: this is essentially the dominant vulnerability that modern aircraft are exposed to. Thus it is crucial to gain systematic insight into the factors and circumstances that drive this kind of three-way breakdown.

Current research

Over the years, the human factors community has tried to gain more controlled access to the processes that govern pilot awareness in automated cockpits and to document systematically the factors and circumstances that sponsor its breakdown. Since the late eighties, investigations have been conducted with a variety of foci and a variety of research strategies. Beginning with field observations of glass cockpit operations, researchers zoomed in until they managed to reliably replicate the kinds of problems encountered during actual line flying in more controlled settings.

Taken together, these findings have contributed to our understanding of individual pilots' awareness of automation and the communication that supports co-ordination in cockpits. Each of the research strategies has managed to highlight only a specific aspect of the three-way co-ordination breakdown, for example the communication between crewmembers or the individual pilot's understanding of the automation and its behaviour. These different foci have to do with the elusive nature of situation awareness as cognitive concept: it cannot be observed directly, so what measures must be used to gain empirical access to it? Given this problem, it is especially difficult to target systematically the problem of three-way co-ordination breakdowns where diverging intentions, actions and feedback from one pilot and the other pilot and the automation all interact—in other words the multiple interleaving events and features that led up to the Cali crash.

Sources of experimental validity

In the remainder of this chapter we review three dominant bodies of experimental and field work on human factors in automated cockpits (Wiener, 1989; Sarter, 1994; Orasanu, 1995) for the light they have shed on three-way co-ordination breakdowns. As is usual in psychological research, the extent to which these streams of research produce systematic findings on the three-way co-ordination breakdown hinges largely on the trade-offs that typically have to be made between sources of experimental validity and reliability in the respective work.

Internal validity derives from the degree to which the experimenter is actually in control over the variance in results. The question is, can the experimenter create and control the circumstances that produce breakdowns in flight crew awareness of automation status and activity? External validity is the extent to which these results are actually generalisable to the situation that the research purports to represent—in this case the three-way co-ordination breakdown. Construct validity is a special form of external validity which relates to the experimental operationalisation of the concept under investigation. In this case, how is "flight crew awareness" operationalised? This also determines (or flows forth from) how it is measured in the studies. Finally, reliability relates to the replicability of results across practitioners and situations. Do the same kinds of circumstances systematically trigger the same kinds of breakdowns in crew awareness? Research on flight crew situation awareness in automated cockpits has juggled these questions in different ways, producing converging evidence on the circumstances and ways in which awareness can break down.

Field studies on crew awareness of automation behaviour

Examining the impact of automation on crew communication and co-ordination was an explicit goal in a series of field studies that Earl Wiener and colleagues conducted in the late eighties (Wiener, 1989; Wiener et al., 1991). Their research methods included interviews with pilots, surveys of pilot attitudes towards automation and field observations of glass cockpit operations, both in line and training situations. Among the findings were that automation tends to induce a breakdown in the traditional roles and duties of the pilot flying versus the pilot not flying. It may enable first officers to make flight path decisions (through the automation) that were traditionally the prerogative of the captain. These characteristics are compounded by the fact that in automated cockpits it is hard to monitor the work of the other pilot. One overall finding was that while highly automated flight decks increase the need for more communication and co-ordination, but also make this co-ordination more difficult.

Field observations are one way to get empirical data about human activity that takes place in complex environments and requires much practitioner expertise to carry out (Woods, 1993; Orasanu & Connoly, 1993). Repeated observations of crew co-ordination in automated cockpits can:

• discover and confirm that it (a particular phenomenon) happens;

- tabulate the factors that seem present and perhaps even contributory in those situations (the independent variables in classic experimental design language);
- begin to sketch patterns for the type of breakdowns that seem to occur over and over. Example: the going sour pattern (Woods & Sarter, 1998), where a triggering event leads to a sequence of misassessments and miscommunications between crew and automation that slowly manages the aircraft into hazard.

Field observations do not demonstrate a researcher's control (and thus a researcher's full understanding) of a phenomenon. They do not allow a researcher to conclude with certainty that the tabulated factors (e.g. not co-ordinating a computer input) indeed determine a significant portion of the variance in observed behaviour. This is the internal validity problem that scares so many experimenters away from field studies in the first place. Given that no certainty exists over all the factors (independent variables) that are necessary to produce the phenomenon (the breakdown in awareness), a researcher cannot make the phenomenon happen. Too many factors that might be relevant or contributory are outside his or her control or even unknown.

But through its inductive and of necessity repetitive nature, field observation brings its own sources of experimental control. Through induction, facts are gathered and then generalised towards theories that capture underlying psychological mechanisms that may be responsible for observed effects. Field observations are directed at the discovery and description of patterns that recur across similar circumstances. In this sense, experimental control does not necessarily represent a missing link in field research at all. And in fact the discovery of patterns allows us to formulate hypotheses—one of the main contributions of exploratory research. Investigations aimed at hypothesis generation form a crucial step in ordering phenomena in the real world and in gaining more controlled empirical access to them.

With respect to the three-way breakdown, however, cockpit field studies to date have not clearly parsed out the causes for breakdowns in flight crew awareness. Although fertile circumstances for breakdown can be recognised and described, the lack of internal validity (certainty of what manipulation or factor in the world causes which effect) limits the ability to draw detailed conclusions about how the interaction between two crewmember's mindsets, expectations, etc. and features of the automation can conspire against flight crew awareness.

Empirical studies with a confederate pilot

The internal validity problem has one clear solution: achieve tighter control over the factors that generate the phenomenon of interest. This is the hallmark of doctoral research done by Nadine Sarter (1994) which has been widely reported (e.g. Sarter & Woods, 1991; Sarter & Woods, 1992; Sarter & Woods, 1994a; Sarter & Woods, 1995; Sarter, 1996; FAA, 1996; Sarter & Woods, 1997; Sarter, Woods & Billings, 1997).

Following surveys that further probed and detailed the earlier Wiener findings, this research conducted full mission simulator experiments with 18 experienced Airbus A320 pilots. The major focus was to explore the nature of and reasons for breakdowns in the co-ordination and communication between pilots and the automation. An important objective of the experiments was to explore whether pilots have difficulties to detect unanticipated changes in the status and behaviour of the automation. This awareness (or, rather, lack of awareness) was inferred from secondary performance measures (for example the aircraft speeding up because of an incorrect automation mode; the pilot himself continuing with an approach which should have been broken off) which indicated whether the individual pilots had or had not picked up on the experimenter-induced change.

An important ingredient in these two-crew cockpit studies was that one crewmember was a confederate pilot: a pilot who knew about the experiment and its goals and played a part in conducting the study. Only one pilot was the subject, the one whose behaviour was of empirical interest. The confederate represents a large source of internal validity. In part through the confederate right there in the cockpit, the experimenter can carefully control the factors that impact the subject pilots' awareness and conclude with certainty that those factors are indeed responsible for any breakdowns observed. What is powerful about this research is that by such control, it shows far-reaching understanding of the phenomenon under investigation. If we can control it, we certainly comprehend it. Pilot awareness, for example, can be eroded by making the automation not show expected behaviour, or not saliently show sub-system failures. What is also powerful is that this research laid bare things that other studies had not dared to touch previously: the large gaps in pilots' mental models of the structure and behaviour of the automated cockpits they fly in daily.

Tight internal control, however, has costs—creating the inevitable research trade-off. In this case the cost was borne by external validity, or more specifically, construct validity. For how was pilot awareness of automation status and activity operationalised? In one of the scenarios, a navigation beacon signal was lost on approach—a reason for immediate go-around. In another, the aircraft stayed in an unexpected mode after go-around, which could result (and did in many cases) in overspeeds. Awareness (or the lack thereof) was inferred from the pilot's lack of reaction and/or the aircraft's continued behaviour along the unintended path. In one sense, what was measured was whether a pilot did or did not see a particular automation change or lack of change. From the three-way co-ordination breakdown perspective, the interesting question to ask is, where was the other pilot? In this case the other pilot was a confederate, and could not—by nature of the experiment—contribute to the subject pilot's awareness of critical changes. In some sense, what was measured was how long the combination of the confederate and the automation could lead the subject pilot into not picking up subtle changes in automation status and behaviour. Not only the automation had to be strong and silent to achieve this; the confederate had to be silent too.

Of course a confederate has only limited impact on external validity. In fact, the survey conducted as part of this research pointed out conclusively (together with similar research) that single-pilot cockpits are in effect often created on automated flight decks, for example, when one pilot is typing the computer and the other slips out of the control loop, or when the other pilot is engrossed in more manual tasks, such as reading charts and setting the aircraft's basic navigation instruments up for an approach (indeed part of the Cali scenario). In these situations, the other crewmember is as good (or as useless) as the silent confederate. But the results achieved in controlled experimental settings with a silent confederate are externally valid only to the extent that they map onto these specific target situations. In other words, their generalisability is limited to these circumstances, independent of how frequently they occur on the line.

The results from these studies are partially applicable to the three-way breakdown. They demonstrate, under tight experimental control, that one pilot can be pushed out of the control loop by silent and dynamic automation if the other pilot is occupied with different things as well. In a limited empirical sense, then, Cali could be replicated. The confederate captain could be asked to deliberately make an erroneous computer input and be silent about it and its results, so as to see when the other pilot will catch on. What these studies don't do is give both pilots the naturalistic chance to develop individually their own mental model of what is going on, against the background of their own expectations, previous assessments and workload— out of which they could exclaim, "hey, do you see what I see?" ("we just lost the NDB signal", or, "we're still in *that* mode"). The issue goes back to construct validity. Awareness of automation activity in automated cockpits is more than a function of one individual keeping his or her eyes open for unsuspected changes against the odds of a silent machine interface and a silent fellow crewmember. It would be interesting to find ways in which we can re-insert this natural collaboration experimentally while retaining tight internal validity. This would allow us to draw conclusions on the factors that drive differential mental models and behaviour on the part of two individual pilots in more naturally occurring situations, i.e. in situations where one of the pilots is not pursuing the researcher's agenda (i.e. is not a confederate), but his own, just like the subject pilot in the research discussed here.

Research on cockpit communication and crew awareness

How aware is a flightcrew jointly of the situation around them? This was the question in studies conducted at NASA Ames in the early nineties (Orasanu, 1995) in which pilots were left to develop their own ideas and models and left to co-ordinate in order to solve problems. The focus was on cockpit teams operating in their natural contexts, and their co-ordination was taken as the probe for their joint awareness of developing problems.

As indicated previously, awareness can only be known indirectly; it has to be inferred from behaviours related to events in specific contexts. As said, the measure used to infer a crew's joint situation awareness in the Ames studies was communication between the human crewmembers, using verbal process tracing (Ericsson & Simon, 1980; Woods, 1993). As in Sarter's studies, flight crews here flew missions fraught with problems and adversities in full-fidelity simulators. Two types of crew utterances were of particular interest—those involved in situation assessment and those related to planning a course of action. Since situation awareness is thought to be at least in part about the future status and behaviour of the system, it was gauged for example from remarks that predicted or alerted to possible future events ("keep your eye on that temperature there").

Among the results were that captains of more effective crews (who made fewer operational or procedural errors) verbalised a greater number of plans than those of lower performing crews and requested and used more information in making their decisions. Joint situation awareness can be said to be achieved through such cross-cockpit co-ordination, where crewmembers support one another in picking up and comprehending the cues and changes around them. This raises interesting questions about whether situation awareness can be improved by teaching specific communication skills or even proceduralising certain communications that would otherwise remain in the realm of relatively unregulated CRM behaviour (some airlines have resorted to such proceduralisation).

Joint situation awareness was measured through communication, but to what extent is communication a measure of awareness? Crewmembers may be aware of much more (even jointly) then they communicate explicitly about. One way to deal with this is to extract problem-related talk from the overall verbal protocol and focus on that. But even then, more effective crews sometimes communicate less, depending on the nature and urgency of the problems afflicting their aircraft (see Wiener, 1993). What was found, however, was that a greater *proportion* of higher-performing captains' (as compared to lower-performing captains') total talk was devoted to flight problems (i.e. statements of goals, plans, and information requests). And this effect was actually greater during high-workload, abnormal phases of flight.

But does a higher level of problem-related communication reflect a higher level of crew situation awareness? Verbalisations appear to be useful for developing shared models for the problems and for assuring co-ordinated actions in dealing with them (Cannon-Bowers, Salas & Converse, 1994; Orasanu, 1994). Certain temporal sequences between events, utterances and actions in this research suggest but do not confirm such a relation. Further, communication may fail to reflect moment-to-moment changes in situation awareness. Presumably it lags behind the event of interest some indeterminate time.

This research was not specifically focused on automation, or how communication reflects awareness of automation status and behaviour. In other words, automation did not enter into these experiments explicitly as a third crewmember with whom co-ordination was also necessary (see Hollnagel, 1998 for a discussion on how to measure cognitive functions (such as automation awareness) in terms of contributions from the entire distributed cognitive system which contains both human and machine agents). To gain systematic access to the three-way co-ordination breakdown, we should make automation a part of the experimental equation. For example, if much interaction with the automation is necessary to cope with certain system malfunctions, how does this influence the overall cockpit co-ordination and performance results?

Construct validity

Still, the overall research problem here is one of construct validity, with or without automation in the equation. Communication (with human or machine crewmembers) is unobtrusive to measure and has high face validity, but as sole indicator of joint awareness it is a limited experimental operationalisation. What we have seen on the other hand is that having a silent second (confederate) crewmember in the cockpit is also a limited operationalisation of automation awareness (i.e. it may be valid only for a certain number of real cases). Automation awareness is more a function of collaboration than this experimental paradigm allows it to be. Yet measuring collaboration alone is a poor indicator of awareness.

Integrating automation and experimental validity

One way out of the experimenter's dilemma is to back up the verbal protocol of activities in a naturalistic, automated cockpit (with two non-confederate crewmembers) with various other datatraces. These traces can help calibrate the extent to which cross-crew co-ordination is representative of awareness of changes in the automation. For example, by tracking head- and eye movements and measuring visual cortical brain activity of individual crewmembers, a researcher can calibrate the lags between annunciation (e.g. of a mode change in the automation), perception, and cross-crew co-ordination.

The lag between annunciation and co-ordination (taken as one measurement unit in the Ames research) could then be filled with another marker: the perception of the annunciation by individual crewmembers. Thus we could gain insight into the different reasons behind the different phases of the lag annunciation—co-ordination ("do you see what I see?"). The lag between annunciation and perception would likely be a joint function of the individual's expectations and properties of the automation feedback (Neisser, 1976). This could help us understand for example why Tafuri does not comment on the turn towards Romeo. There is nothing in his mental model that helps him expect such behaviour and subsequently—busy as he is with other things—he is not looking for evidence of it either. The second lag, between perception and co-ordination, would shed light on different processes altogether: This second lag has to do with the observer's expectations of what the other pilot will have told or will expect the machine to do, what the other pilot is busy with at that moment, and any other assumptions or observations about the other pilot. This second lag can help explain Williams' silence about the aircraft's behaviour: the turn is consistent with his mental model and assumed to be consistent with Tafuri's mental model as well. That is why Williams doesn't mention anything.

Of course, even with eye tracking and the measurement of visual brain activity, an experimenter can't be sure what meaning the crewmembers endow an annunciation with (for example, it says V/S, but it is still interpreted as FPA)—only that a certain change was perceived. There are no inherent ways to measure what sense the observer makes of the data. In order to get access to the interpretation, a researcher would need to carefully monitor and code subsequent crew behaviour, or even resort to post-session debriefings to help explain their interpretations.

Conclusion

The glass-cockpit procedure to co-ordinate computer entries before they are executed is not consistently followed. Research to date has identified multiple possible explanations that rely on features of the tasks, tools and goals that human pilots have at the time. However, such reasons are unable to capture the more subtle breakdowns in three-way co-ordination (between two pilots and the automation) that are for example observed in the sequence of events leading up to the 1995 Cali crash. Also, such explanations are not able take possible countermeasures beyond reinforcing the procedure to co-ordinate entries —especially because of, or despite, circumstances conspiring against it. This chapter has reviewed some existing research on crew awareness in (glass) cockpits and laid out how trade-offs in sources of experimental controllability and validity require us to search for new approaches that can push the research frontier into systematic empirical studies of three-way co-ordination breakdowns between two crewmembers and their automated cockpit partner.

6 Filling the Gaps in the Human Factors Certification Net

GIDEON SINGER

Saab Commercial Aircraft, Sweden

Introduction

In most commercial modern aircraft today navigation and flight path are controlled by a system called the *Flight Management System* (FMS). The interface with the crew in the cockpit is achieved by an active display that is usually called the *Control Display Unit* (CDU) and is the input/output channel. In addition, the system displays its planned/executed route on a *Navigation Display* (ND) and flight path guidance by means of a Flight Director on the *Primary Flight Displays* (PFD).

The technical requirements and methods for testing system performance, accuracy and fault analysis are well established and are defined in the certification requirements for such systems and aircraft. The methods for technical validation of such systems have been used successfully in many types of aircraft avionics and have shown very high level of reliability. However, validation of the way the system interfaces with the crew, displays information or reacts to crew inputs is not well defined and each manufacturer or vendor is free to adapt its own philosophy and methods of showing compliance.

The FMS was introduced into the transport category aircraft as an integral part of the cockpit and was evaluated and certified as a novel system by way of extensive simulations and crew evaluations. Aircraft of the Boeing-757 and Airbus-A320 generation were the first to have advanced integrated cockpits with a crew of only two pilots. The two crew cockpit was evaluated in length by manufacturers, airlines and authorities in regarding to crew workload when using systems such as the FMS. In the late nineties, new ATC regulations that mandate the installation of FMS into *all* aircraft that use medium and upper airspace in Europe were published. This created a growing demand for cheap "add-on", "Off-the-Shelf" systems for integration in older aircraft of all sizes.

The majority of the transport category aircraft flying today have not been designed with FMS as part of the cockpit. Many aircraft are still featuring the so-called "Classic Cockpit" and are lacking the modern avionics bus communication technique used on the integrated cockpits. The majority of the aircraft requiring new installations are of the commuter/regional type, which are typically turboprop aircraft seating 19-70 passengers. In many cases the installation is made by the operator as a minimum-cost solution to comply with the operational requirements. A thorough evaluation is often not done, nor are these minimum-cost solutions guided by human factors approach, as was the case with the integrated FMS designs. When such designs are evaluated, it is for the *correct function* of the system during normal operation but not for *possible error modes* and *crew mismanagement*.

Different FMS vendors have developed and evolved different system philosophies and interface logic. In many cases these were based on a large project that dictated certain features to interface with the rest of the cockpit (especially the electronic navigation displays). In many of today's "add-on" installations, the design process is reduced to the functional aspects. Crew feedback and interface are often deficient and seldom tested or validated. In addition, many features that were active in an integral system (CDU, ND and other displays) are missing since the aircraft is often of an older generation displays and lacks the fast avionics data busses.

In the 1996 FAA report and other published incident databases, the category of aircraft with "Add on" FMS systems are usually not included. With regards to the high percentage of Human Factor errors in operating the FMS in the aircraft with integrated cockpits, it is of great concern for even higher error rates in the "add-on" designs.

Purpose of human factors certification

Human Error in the integrated cockpit has been found to be one of the main causes contributing to incidents and accidents in the past decade. The FMS interface, being the most complex interface in the cockpit, has been shown to be one of the weakest links in high workload, dynamic situations in flight (Billings, 1997, FAA, 1996, Woods, 1998). This trend is probably going to increase with the introduction of many older aircraft with "add-on" solutions that have only gone through a minimum effort approval process. Both integrated and "add-on" installations lack an objective evaluation and validation methods for the system in its correct operational environment. This results in an uncertainty for the manufacturers, operators and aviation authorities as to the predicted level of safety achieved by the crew operating the systems.

Clearer methods and criteria are needed in order to quantify the deficiencies in human factors design and be able to mandate an improvement at an acceptable price. As long as such requirements are not available (even as advisory material) changes will be deferred to the future and the potential for error will prevail.

This chapter addresses some of the FMS interface deficiencies, explain the potential for error and the possible effects of such an error. It will then show the limitations of today's certification requirements in regulating such deficiencies and suggest new requirements for future use. An attempt will be made to give specific guidelines with a Pass/Fail criteria without mandating a specific system architecture or logic. The main goal is to highlight the need for an objective validation method with sufficient statistical value that will allow the system to stand up to a Pass/Fail criteria before entering service. This evaluation method, that has been used successfully in a different context, will be explained and its merits and limitations discussed. The proposed criteria will be in a format that will allow an aircraft manufacturer, airline or *Supplementary Type Certificate* (STC) applicant to evaluate a new FMS integration in a cockpit with a limited effort and acceptable cost. As future Advisory Material this method could be used by the different aviation authorities to form a standard basis for acceptance of new and retrofitted systems, and to aid the evaluating teams during the certification process.

This chapter will be limited to the basic function of a FMS, called *Lateral Navigation*, which is the function mandated by ATC and used for navigation guidance in the horizontal plain. Lateral Navigation functions are available on all FMS makes and are relevant for all aircraft types.

Present certification requirements

Any aircraft in service must meet certain rules and regulations set by the authorities. Due to the public interest it is the large aircraft used for paying passengers that have to meet the strictest rules within the civil aviation world. These rules were initially set and standardised in The Chicago Convention and formed the basis for the International Civil Aviation Organisation (ICAO). ICAO appendices address all aspects of air transport and are the basis for national and regional requirements set by each member country. In the western world today, most countries follow the rules set by the American Federal Aviation Administration (FAA), the Joint Airworthiness Authorities (JAA) for most European countries, or close derivatives of such rules.

The rules that are of relevance to this paper are ones addressing the design process of an aircraft and the way it is used in service. Both the design requirements (also called Airworthiness requirements) and the rules of how an airline must use the aircraft (called Operational Requirements) must be fulfilled in order to legally operate commercial aircraft.

The Airworthiness rules see to it that the aircraft will be designed to correct standards, meet system safety levels and meet minimum levels of control in all foreseeable conditions. In the case of large commercial aircraft (above 5.7 tons) the set of requirements are called Federal Aviation Regulations Part 25 (FAR 25) or the Joint Airworthiness Requirements (JAR 25).

The fact that an aircraft is built to the airworthiness standards does not clear it for operations. Separate requirements exist for each land and type of operations that mandate the rules set for the airline regarding aircraft equipment, it's use, crew training, duty time, maintenance etc. In the USA, operational rules are regulated by FAR 121 or FAR 135 and in Europe JAR-OPS. Other more specific requirements exist to regulate more specific areas of operations (JAR-FCL for aircrew certificates, JAR-HUDS for Head Up Display operations etc.).

The mandatory FAR/JAR 25 paragraphs are usually very generic regarding new technologies. The FARs are not always identical to the JARs and at times result in conflicting requirements. It is only lately that an effort has been made to harmonise (publish a common rule) the two sets of requirements. In order to aid the applicants in complying with the requirements many detailed advisories have been published during the years. The Advisory Material (AC and ACJ) are more detailed but even they do not address the human factors issues fully.

As new control systems and avionics suites are in demand, the applicant for a new system is usually faced with a long and complicated process of attaining *Special Conditions* that complement or override existing regulations. (One such example was the approval of the AIRBUS 320 flight control system in the USA).

When applying for approval of a new FMS installation, whether on a new or an existing aircraft, several rules or paragraphs in the aviation "Law Book" must be addressed and compliance must be shown. Many of the technical requirements exist for system design but only the following design requirements are mandatory for compliance regarding Human Factors when applying for a FMS certification:

Mandatory Requirements:

- **FAR/JAR 25.671(a):** *Each control and control system must operate with the ease smoothness and positiveness appropriate to its functions.*
- **FAR/JAR 25.771(a):** *Each pilot compartment and its equipment must allow the minimum flight crew to perform their duties without unreasonable concentration or fatigue.*
- **FAR/JAR 25.777(a):** *Each cockpit control must be located to provide convenient operation and to prevent confusion and inadvertent operation.*
- **FAR/JAR 25.1301(b):** *Each item of installed equipment must be labelled as to its identification, function or operating limitations, or any applicable combination of these factors.*
  (d) …function properly when installed.
- **FAR/JAR 25.1329(f):** *The system must be designed and adjusted so that, within the range of adjustment available to the human pilot it cannot produce hazardous loads on the airplane, or create hazardous deviations in the flight path, under any condition of flight appropriate to its use, either during normal operation or in the event of a malfunction, assuming that corrective action begins within a reasonable period of time .*
- **FAR/JAR 25.1523(a):** *The minimum flight crew must be established so that it is sufficient for safe operation considering the workload on individual crew members (see Appendix D).*
- **FAR 25 appendix D:** *Criteria for determining minimum flight crew (a)(3) Basic workload functions … Navigation. (b) Workload factors. The following workload factors are considered significant when analysing and demonstrating workload for minimum crew determination: (1) The accessibility ease and simplicity of operation of all necessary flight, power, and equipment controls…(8)The communications and navigation workload.*

Guidelines:

- **AC 25.15:** *Approval of Flight Management Systems in Transport Category Airplane - (Addresses mainly vertical navigation, mode presentation and go-around logic).*
- **AC 25.1309 (1a)** *System design and analysis includes…effect of system failures on flight crew workload.*

To the reader not familiar with the requirements or the aviation domain, some mode detail is needed here. By explaining each regulation and what it actually requires of the designer or operator it will become clear what the requirements are missing in the Human Factors aspect.

In **FAR/JAR 25.671(a)** the issue of cockpit controls and their ease of operation is addressed. The terms *ease, smoothness* and *positiveness* are actually inherited from the "Classic" cockpit containing only levers, switches and buttons. In the domain of electronic screens and keyboards this requirement is very vague since feedback is not always felt in the control itself (keyboard buttons). How would you translate these terms into software dialog on a modern display?

The **FAR/JAR 25.771(a)** addresses the *concentration* and *fatigue* issues when working in a cockpit. This paragraph, when used for physical strain of operating the controls in an older cockpit, was reasonable and easy to show compliance with. Since modern cockpits require almost no physical effort, the means of measuring mental fatigue are not described and neither are the acceptable levels stated. How should one use this requirement to measure an acceptable level of concentration of fatigue when battling against a stubborn FMS during a late runway change in difficult flying conditions?

Even the **FAR/JAR 25.777(a)** is mainly addressing levers and switches so that they are designed to *prevent confusion and inadvertent operation*. This is easily tested by abusing the use of the levers (like landing gear handle) prior to approval. When implementing the same requirement to the FMS control panel (CDU), this requirement cannot be tested due to the infinite number of possibilities and probably impossible to achieve (since *prevent* is an absolute term). When using the CDU in turbulent flight conditions, even simple slips are unavoidable especially if the screen is of a Touch Screen type. Shouldn't terms like *probability* and *effect* of error be included in the requirement? Terms like *Fault tolerant*, and *Undo functions* are missing and thus not required by the authorities.

FAR/JAR 25.1301(b) is a straight forward rule regarding normal cockpit controls (or is it? Does the control column have "Up" and "Down" labels on it?). How do you define *labels* on a display? When checking existing displays it becomes clear that there is no standard and that more guidance is needed. Even here, terms like *effect of error* should have been used. The requirement of *function properly when installed* is easy to evaluate on mechanical levers, but how should that be done to an acceptable level of certainty on a CDU of a modern FMS? How should one treat underlying menus and functions and the acceptable level of access of the operator to that display layer? In the PC environment of the new home computers a certain level of display architecture has become a standard and is addressed in the International Standard ISO 9241 guidelines. Shouldn't such guidelines be made for flight critical systems like the FMS?

FAR/JAR 25.1329(f): is the first paragraph quantifying the risk level acceptable following an error. It states that pilot input may not *create hazardous deviations in the flight path*. This can be translated to the system being allowed to deviate and cause damage and limited number of deaths onboard every 10,000,000 flight hours (for the whole fleet!). How can this be tested in advance? How many hazardous deviations per year are acceptable? (If one assumes 10,000 aircraft in the large aircraft world fleet, one could expect up to 20 limited accidents yearly and still meet this requirement). How do we address pilot entry of ambiguous way-points in the database (more than one way-point defined by the same identifier/name)?

FAR/JAR 25.1523(a) and **FAR 25 appendix D** are the paragraphs that specifically address *workload on individual crew members* which is a very good way of testing for appropriate human factor consideration in the design. These rules were used for determining the ability of a crew of two pilots to handle the advanced cockpits without the aid of a flight engineer in the proof of concept in the early eighties. There is no method or criteria for measuring and finding acceptable levels of workload especially in the *communications and navigation domain* when using the FMS. Today, the default crew is two pilots on all modern aircraft and applicants are no longer required to perform lengthy tests proving the concept. Another issue is the *ease and simplicity* of system use required. How do we determine what is simplicity and how much training is needed for the crew to reach the level of ease stated above?

AC 25-15 and **AC 25.1309 (1a)** are more detailed guidance paragraphs. The first explains the functions and logic that are required for the FMS integration in order to meet required aircraft performance. The second paragraph gives guidance as how system error is taken into account when analysing total system safety. It actually addresses *effect of system failures on flight crew workload* but treats only component technical failures and not crew error in using the interface. One could claim though, that if paragraph 25.777 *prevents* **all** errors, then this becomes a non-issue, but it is clear that this could not be the case. How does one measure crew workload and in which environment and with what crew?

So what is missing in today's requirements?

It becomes clear that the existing requirements are general, vague and do not aid the manufacturer or authority evaluation groups in their decisions. Even if a reviewer sets a requirement for a workload experiment with the system, he would have difficulties in substantiating the request against cost and schedule constraints.

None of the above paragraphs include any method or criteria for compliance. When evaluating component technical reliability and failure modes it is required to show detailed analysis or experiment reports of system behaviour are the most extreme conditions. For critical components in flight controls and engine controls the component is abused for long periods of time in order to detect latent failures (so-called "Shake and Bake" testing).

The nature of the requirements results often in reviewer statements of a successful result that are just as vague and generic since they mimic the requirement text to the letter. The statement is very subjective and usually based on a very limited exposure to the system in very few operational conditions. *Workload, fatigue, ease* and *simplicity* are meaningless terms without a frame of reference to them. An example of such a review could be one of a new FMS in an existing aircraft. The reviewing pilot's previous skills, background and exposure are not questioned. As a reviewing pilot he/she is always very experienced and involved in previous development and testing of the system. The reviewers are always biased by other factors that affect their judgement and they do not represent the average user's background in service. Such reviews are productive during the development phases of a system and as a complementary expert opinion, but could be misleading as compliance documents. The pilot statement would probably say: *The new system was easy to use and did not result in excessive workload or fatigue in all flight conditions. This shows compliance with paragraphs ....* For anybody engaged in experimental methodology it becomes clear that the validity of such a statement could be questioned for many reasons.

Guidance from existing research

The introduction of sophisticated automation for the control of flight path and navigation has resulted in increased safety, reliability, economy and comfort in the last 20 years (Wiener & Curry, 1980). The trend in safety improvements presented by Boeing (Billings, 1997, p.182) is clear especially in the first few years, and Airbus shows a continuos improvement since introducing the 3rd generation aircraft such as the A320 (La Burthe, 1996 presentation to the FAA). The advantages in reliability are mainly due to the new generation electronics that have lower failure rates and are easier to troubleshoot and replace.

Economy and comfort have been improved by reducing weather minima for landing, and by allowing direct routing between destinations which saves time and fuel. Indirectly, the new automation has allowed the decommissioning of conventional land based navigation aids, thus saving money for the tax payer (Billings, 1997).

Increased automation in transport aircraft has improved the technical reliability and flight safety but accidents have happened even with new generation aircraft. The number of accidents per passenger mile was once decreasing but since the volume of commercial flights has constantly been on the increase, the level of accidents has become alarming. The 1996 FAA *Human Factors Study Team Report on the interface between flight crews and modern flight deck systems* cited that about 70% of the fatal accidents involving new generation aircraft are due to Human Error.

A similar conclusion to the one made in the FAA report above, was made earlier by the ATA *1989 Human Factors Task Force report*. It questioned whether the new automation was fully compatible with the capabilities and limitations of the human in the system (Billings, 1997).

The 1996 FAA report found the FMS to be one of the important contributors to Human Error and loss of situation awareness. Since the FMS is now integrated into the aircraft's flight path control (Lateral and vertical navigation) and other performance related aspects of flight (weight, takeoff speeds, engine power and fuel calculations) such errors can lead to catastrophic effects (Cali - B757, KAL shootdown - B747, Collision Delta L-1011 & Continental B-747).

Sarter & Woods have written several papers addressing the FMS automation problem. In their latest paper, *Learning from Automation Surprises and "Going Sour" Accidents (1998)*, they summarise the research effort of the last years in the field, much of which relates to the FMS. The paper criticises new designs for their *Technology-Induced Complexity* and the way training tries to compensate for these deficiencies. The paper also quotes designers excuses for system deficiencies. When coming to suggestions for improvements, the authors give only general theories of how to improve the design. Terms like *improved feedback, problem-driven design,* and *activity-centred* or *context-bound* design are mentioned without going into a constructive methodology for the implementation of such ideas. When addressing the problem of *Display Observability* the authors state their opinion based on several subjective pilot comments instead of basing their claim on experimental results from realistic environment (simulators).

Jens Rasmussen, in his paper *Ecological Interface Design for Reliable Human-Machine System (1998)* addresses the problem of modelling the complex human-machine in a new light that explains several of the deficiencies in today systems. Adaptive behaviour in a human-machine system points to the need for higher level abstractions rather than the usual modelling in terms of events, decisions and errors. Rasmussen explains that in order to evaluate the reliability and functionality of a complex system we need to judge whether it will work during an extended period of time when adaptation has taken place.

In order to address the above research results and recommendations, this paper will try to provide an idea for implementing a method that will identify system human factor deficiencies based on an extended period of use in a realistic environment and with objective data as an output.

Today's certification status

Aircraft design, development and certification has evolved during the years into a well structured and thorough process. When an aircraft manufacturer has installed a new system like the FMS into a new or existing aircraft design, the following *Airworthiness Approval* process is usually being followed.

An application is made to the certification authorities in the form of a detailed *Test Description*. This description defines the new system functions, explains which specific design requirements it will comply with and lays out how compliance will be shown. Compliance may be shown by means of simulation, flight test or pilot evaluation.

The system goes through a phase of *Development Testing* which is an iterative process. Here changes are made to optimise the system for its intended use. In this process, test pilots end engineers test functions, logic, display and interface with the other aircraft systems on several simulations or test flights.

When the test team is satisfied with the results it is time to "sell" the evaluation process to the aviation authorities for them to test and approve the system. *Certification Test Plan* is the document of the agreed upon process that must be presented to the aviation authorities. This testing is then performed by either the authority evaluation team or is delegated to the manufacturer test team. All data is recorded and used for reporting the results.

Fit, Form and Function are evaluated by several Test Pilots engaged in the development phase. This includes the ergonomics of the interface units, interface with other controls in the cockpit and readability in different lighting conditions. This process is very subjective and is based on the experience of the reviewing pilots. This process is completed with a report written by the Design Reviewer (a senior test pilot) which states that the system has been evaluated and found not to affect any other systems. This report does not show specific compliance with any regulatory requirements but is mainly for the internal process of quality control of cockpit changes.

A *Certification Test Report* is then produced based on the previously agreed upon Test Plan described above. This report shows how each item has been tested successfully and is the basis for the approval given by the aviation airworthiness authorities. Most system performance and reliability results must comply with objective guidelines and are tested and reported in such a manner. The means of showing compliance can be either flight testing the system in a test aircraft, testing it in a flight simulator (for dangerous flight conditions) or by component lab testing. In some cases, like when such a system has been approved on another similar aircraft model, similarity with previous designs and theoretical analysis may be used as the sole means of showing compliance.

Crew Evaluation Certification Reports are used to cover all items that the engineers do not have test or analysis results for. These are all the qualitative requirements that are stated in the requirements and that relate to the way the pilot communicates with the system. Issues such as ease of *readability in flight, tendency to mislead* and the need for *excessive skill* of the operator are covered by these *Crew Evaluation* statements written by a senior pilot and approved by a reviewing pilot.

When all the above mentioned documents have been approved the system is declared *Airworthy*, meaning that the installation in this specific aircraft has been found safe for use.

In order to approve the system to be used in that aircraft when carrying passengers (or cargo), each operator (airline) has to pass an *Operational Approval* that includes the following:

♦ Documentation of how the new system will be integrated into airline procedures is to be approved by the local operational authorities (The operational authorities are not the ones involved in the airworthiness approval).

♦ The airline must submit a full training plan for pilots and maintenance personnel including detailed simulator training syllabus and aircraft system monitoring programs.

♦ In case of flight-critical systems (requiring the highest level of safety and reliability), a track record for all aircraft and pilots in the fleet is required. This track record states the number of flights each aircraft/crew used the system and whether the process was successful. This assures a minimum exposure to all aircraft and pilots prior to allowing the crew to fully rely on it. If during this process any particular risk for crew error is identified, the operator is required to introduce *limiting procedures* in order to add safety margins for operations (like better weather requirements for landing while using the system).

As shown above, the airworthiness and operational approval process is covered by detailed requirements and compliance documents. However, the human factors aspects lack any objective criteria with Pass/Fail levels. Due to the lack of objective evaluation methods in this process, most systems introduced into service contain deficiencies that increase the risk for crew error resulting in unknown effects.

Deficiencies in current designs

In order to explain the problem in more detail, examples from present modern designs will be described and deficiencies highlighted. The risk involved in these features will be shown and possible effects of making an error will be discussed.

CDU design

The CDU is the main interactive interface between the crew and the FMS. Some newer designs include cursor and mouse-pad features (Boeing 777) but these features are not yet considered as standard and will not be addressed here. The standard interface includes a screen, line select keys, alpha numeric keys, specific function keys and warning lights. This design is typical to all manufacturers, but unlike the typewriter, **there is no standard position for the letter keys**. In addition, each specific function key has a different use in each design. The NAV key for example; on one design is used for *navigating* while on another design it is used for *planning* only. The FPLN (Flight Plan) key; on one design used for *planning* while on the other used for *navigating*. These features increase the risk of *negative transfer* when transitioning between systems.

CDU menu layers

The CDU has a limited display area (usually capable of 8-10 rows of text) and in order to integrate all the functions required for navigation and performance, several menu layers are needed. The limited display area, when interfacing with large size navigation and performance data required for today's airspace, emphasise the Keyhole Effect (Woods, 1984). This effect is the property of having a small viewport relative to the large size data it is used to interface with.

Since layers are unavoidable, the design aim should be to minimise the number of layers and more importantly, to *keep the operator oriented at all times* as to his position and his possible hidden options.

On the successful designs, the interface is such that menus and sub menus are organised in a manner that each layer includes line selectable cues for the *next, previous* and *main* menus. On other designs, in order to incorporate as many functions as possible into the system, each menu can have up to 5 pages or more, each including up to 8 selectable functions with a multiple page menu each. The fact that the positioning of these functions is not logical to the operator means that the pilot needs to search by means of "trial and error" through up to 10-15 menus until finding the right function. Since no cues are available for this search the pilot has to rely on memory or perform the time consuming search. This is a characteristic that increases workload and frustration and should not be a feature in a modern interactive display (Molich & Nielsen, 1990).

One example of a hidden function that is essential for safe operations is the function of transferring a change from one CDU to the other on a DUAL FMS system. On one design this cue is not apparent until the pilot has paged through the entire flight-plan which might require 3-6 inputs.

Feedback on changes

For flight critical actions, like changing flight plan or selecting a direct route to a new waypoint, some designs allow the pilot to *review* the navigational changes and then *accept* or *cancel* it before the change is *implemented*. Other designs are much less tolerant and once a change is initially prompted it is immediately implemented. *Direct To* and *Delete* functions are the normal methods for the pilot to modify a flight plan in flight. These modifications affect the aircraft flight path when implemented and therefore must include cues for crew verification. Since slips and errors are quite common in the cockpit environment due to turbulence, parallax errors or procedural errors, it is essential to have a review function for critical changes. The system in this case must be fault tolerant and include clear review functions.

Display on ND

The CDU, being an alpha numeric display, is not the most optimal display for building pilot situation awareness of aircraft position, trend and navigational forecast. It has therefore become an unwritten standard in industry for navigation track to be depicted on a graphical display (ND or other Multi-Function Displays (MFD)) for both pilots to evaluate. All designs translate the changes made on the CDU to the active track on the ND. When making a change such as *Direct -To* or *Delete,* some designs depict both the active track and suggested change for the pilot to review prior to activation. This display on the ND gives the pilot a graphical depiction of the text change in form of a temporary route change overlaying the active route. This feature allows all crew members to review the change before activation and has been found to increase system fault tolerance.

Database ambiguity

Today's database of waypoints, navigation aids and procedures is enormous. The same identification in the database may define several waypoints around the world. Usually the risk for selecting the wrong waypoint is mitigated by suggesting the nearest one at the top of the list. NDB and Marker beacons have a *two letter code* and can therefore describe several beacons. On some designs the NDB code is not accepted without a dedicated suffix to minimise error, while on other designs an input without the suffix will be accepted without feedback to the pilot that another (non-NDB) waypoint has been selected. With increased database complexity and error potential it is essential to insert reasonability check functions to minimise the risk for selecting the wrong waypoints.

UNDO functions

Any user of a PC software expects an *UNDO* function in order to recall inputs, whether they are due to errors or slips. This function is not a standard feature in CDU software and each design has its own criteria for providing an *UNDO* feature. This feature may be lacking even in the flight critical parameters such as *Direct-To* and *Delete.* In some cases the selection of *Delete* to the wrong line select key cancels the whole flight plan without any *Review* or *UNDO* options.

Colour logic between CDU and ND

In today's cockpits, interfaces between EFIS and FMS of different makes are becoming more common. This may highlight differences in communication standards with the effect that the different displays show different colours for the same waypoint. Combine this with aircraft manufacturers who have different colour standards, and the resulting display might be confusing and misleading. The lack of industry standards for colour coding makes the task of integrating avionics difficult and usually with unexpected results.

Evaluating the risk levels

As in any airworthiness certification process on aircraft, it is essential to determine the effects of each failure mode (FMEA) in each flight condition and decide if the combination of probability and effect is acceptable. This process is well defined when systems are checked in isolation but is not structured enough when the human operator is involved. Human Error has not been given a probability figure except for some classic issues like the omission of critical actions (Landing Gear) where the probability is assumed to be one.

In order to make the design and approval process acceptable in terms of time and funds it is essential to find a method of evaluating the effects of Human Error in systems such as the FMS and address the ones where the risk level is unacceptable.

System component reliability is determined on the basis of past experience. Could Human Error be based on past experience using equivalent systems? How do cockpit procedures and cross-crew monitoring come in as a factor?

Due to the dynamic nature of crew operations it is essential to use a realistic environment for this evaluation. Crew training is performed daily all over the world, using state of the art FMS designs in modern simulators. Several databases have been collated in previous research (NASA, ALPA) that can give a value in the form of errors per flight hour. These values could vary as a function of the design or training and a measurable safe design be chosen as the reference for future evaluations.

In addition, the lack of standardisation between designs increases the risks of negative transfer when transitioning between aircraft with different systems.

Suggestions for new certification requirements

The FAA and JAA Human Factor working groups in co-ordination with the manufacturers (AECMA and AIA) have been trying to produce new airworthiness certification requirements for the future. Unfortunately, these new requirements are being opposed by an industry that claims that training and procedure form more effective methods of decreasing Human Error related accident rates. In addition, the new rules will only apply to new aircraft types that will apply for a new certification basis, i.e. post B777 and A340. Since new FMS installations are relevant even on existing aircraft, the new design certification rules might be valid for future STC approvals or as retroactive Advisory Directives (AD) following recommendations of accident investigations (e.g. NTSB).

When looking at work done in Human-Computer interfaces, we can find very clear guidance of how to improve displays. The International Standards ISO 9241, give detailed guidance as to good practice in computer display design and logic in visual display terminals (VDT) for office work. In Molich & Nielsen's paper (1990), the basics for good display interface are described based on a large study of users. In a form of a checklist, the items of greatest importance for a good display design were found to be:

- Simple and Natural Dialogue.
- Speak the User's language.
- Minimise the User's Memory Load.
- Be Consistent.
- Provide Feedback.
- Provide Clearly Marked Exits.
- Provide Shortcuts.
- Provide Good Error Messages.
- Error Prevention.

The FMS CDU is a very task oriented display with limited screen surface and specific functions. Still, the above recommendations should be applicable even for the FMS interface. In order to achieve an improvement in interface design of a FMS installation, the following, more specific suggestions, are proposed as advisory material for relevant FAR/JAR paragraphs.

In the design of cockpit control interface with the Flight Management System the following rules should be applied:

1. It should always be visible to the operator which MENU is displayed and what are the available options for direct transit.
2. For changes of active flight path (DIR, DELETE, HOLD, OFFSET, INSERT etc.) the change may not be activated prior to a graphical review of the changes (on a 2D display) and a confirmation by the pilot.
3. An UNDO function must always be available and a prompt visible for immediate activation.
4. Not more than ONE page change should be needed for accessing functions for changing active flight path. (ex: DIR, HOLD, DELETE, OFFSET, INSERT).
5. If more than one CDU is available, the data transfer function must be available on each page, or a prompt must be visible without the need for page change.
6. When inserting a waypoint name that has several possible locations, in addition to displaying all options in a logical order (closest first) the system should display a question requiring confirmation by the pilot. This question should include the full name of the selected waypoint.
7. When the system does not accept a value or input it should display the reason for the denial in clear text. This is to expedite error analysis by the operator.
8. When colour coding is used for waypoints on the interactive display (CDU) the same colours should be displayed on the graphical navigation displays.
9. Since CDUs may be installed in different positions and angles relative to the pilot reference eye point, a means of aligning the display to the line select keys should be available.

Suggested methods of testing/validating

Head-up display validation method

The Head-up display (HUD) systems for low visibility approaches has been one of the recent systems that has gone through a thorough certification process that did address the human operator successfully. The pilot was put in the loop, both as a human servo and as a decision maker, in a critical system with new and novel means of guidance. In order to prove that the reliability of such a concept was equal or better than the *Autoland systems* already in use, new requirements were set, some of which addressed the crews' ability to handle the system interface, display and guidance. Unlike existing human factors paragraphs, the HUD requirements were more of an objective criteria than before. In this case, the certification base was not the airworthiness (FAR/JAR25) but rather the Operational Requirements for all weather operation (JAR-AWO and FAR-120.28C(D)). These requirements were formulated due to the need for a quantitative means of measuring performance of a human servo/decision-maker.

The FAA advisory circular AC120-28D (draft) shows in detail how the performance of the HUD system should be evaluated with regard to workload. Paragraph 7.1 states the following:

> "For fail passive rollout systems with command guidance it shall be demonstrated that a safe rollout can be achieved with a satisfactory level of workload and pilot compensation following a failure, using the FAA Handling Quality Rating Scale (HQRS) found in AC25-7.

> For the evaluation of landing systems with manual control and command guidance (HUD), subject pilots must have relevant variability of experience (e.g. number of subjects, Captain /FOs, experience in type). Subject pilots must not have special experience that invalidates the test (e.g. not special recent training to cope with failures, beyond what a line pilot would be expected to have, etc.) Failure cases must be spontaneous and unexpected on the subject's part.

> For initial certification of a landing and rollout system comprised of manual control and command guidance (HUD) in a new type airplane, at least 1,000 simulated landings and at least 100 actual landings will be necessary. For evaluation of these systems, *individual pilot performance should also be considered as a variable* affecting performance. Pilots of varying background and experience level should be used in flight and simulation programs. They should have current licenses and be given training in the use of the landing system similar to that for the line pilots. No pilot should make more than 5 consecutive landings without a break of at least an hour."

The circular specifies, in quite a high level of detail, the requirement for the evaluation. Even this level was not sufficient and was complimented by a written agreement regarding number of pilots, number of runs, airports etc. between the applicant and the authorities prior to the evaluation. The high level of detail specified in the advisory might seem as a constraint on the applicant but in fact it gives clear rules for everyone to follow.

This evaluation process was found sufficient to show compliance with system performance and reliability. But more importantly, it proved that the man machine interface did not conceal any errors with hazardous effects. In addition to showing lack of errors of great consequence it also proved statistically that minor crew errors resulting in a failed approach did not exceed the 5% level (the well accepted success level of 95%).

This method of certification combines compliance with airworthiness and operational requirements and is probably the only objective method of showing compliance with an acceptable Human Factors design. This method ensures that the typical operator, and not only the Test Pilot population, is represented. It enables the evaluator to collect *objective performance* and *error rate data* and evaluate corrective procedures or design issues in order to rectify unacceptable results.

The extent of the testing can be adopted to the criticality of each system. It has been established during the HUD certification process for the JAA that a set of 300-500 simulator landing runs with 5-10 pilots could achieve these goals for a highly critical landing guidance system. The method used for showing statistical convergence during the HUD JAA certification was Sequential Analysis (Ref. Flight Dynamics, Portland, Oregon) which was used on other HUD certifications but has not yet been approved for FAA certifications. To evaluate changes in approved designs, a smaller set may be sufficient.

Suggested FMS validation method

In the case of FMS, the approach and departure scenarios will be more appropriate. Each approach or departure can be divided into small segments that require the crew to use the FMS for making changes. Each segment can be evaluated and compared with an optimal flight path. Random changes in flight plan or clearance will be fed to the crew and their reaction checked from an ATC and terrain perspectives.

In addition to performance and system reliability, Human Factors aspects could be evaluated in the following way:

♦ Non-recoverable crew error leading to reduced safety (altitude busts, lateral separation violations on parallel runways, or reduced terrain clearance) will result in a failure to meet certification.
♦ Recoverable minor errors (affecting only crew response time) will be allowed up to 5% of the runs.
♦ Changes may be made to the system *hardware, software* or *operating procedures* of the crew and the test redone.

This whole test should be done in a full flight simulator for the relevant type of aircraft with the new system integrated to an acceptable level (minor deviations cannot be avoided). The simulator must show a high fidelity for the relevant flight phase and system integration must be correctly integrated. In the FMS certification process simulator flight characteristics are not as critical as for a HUD system but system integration is critical for crew normal and abnormal procedures and situation awareness.

Means of registration must be available for aircraft flight path (standard simulator equipment) and of crew cockpit activities (video or observer).

Expected side effects

In the HUD certification, the above described validation method was mandatory for certification and required large resources for preparation and execution. The resources required were in form of engineering, testing and simulator expertise. The actual validation period was between 5-10 simulator days (ca 100 hours). In addition, the participation of 10 pilots affects the cost and complexity of such a process. Its is quite clear that some objections will be made to such a process for system certification due to the increases cost and time requirements.

In the case of system evaluation, like a new FMS, the scenario does not require the same flight path accuracy as in the case of the HUD precision approach and landing, since the system is not used for a precision task like a landing. In addition, each run can be divided into several shortened sub-tasks, which compresses the crew exposure to the system into a limited time. The cost of pilot participation can be reduced by counting these simulator flights as part of the training required for their future rating on the system.

It is essential that the criteria for pass/fail be accepted by manufacturer, operator and certification authority. It is therefore critical for such a process to have a written basis in the form of advisory material or operational requirements.

Conclusions

The Design and Certification process of new avionics systems in a Transport category aircraft is a long and costly process. The present certification requirements and guidance material do not present any objective means of validating a new design for its Human Factors acceptability in service. Many statements are made as to the acceptability of the system and its compliance with the requirements but none are based on an approved method. This chapter has tried to supply some advisory material for characteristics required in a CDU for it to comply with the existing generic requirements. Based on the previous certification process of the HUD, a validation method has been proposed with Fail/Pass criteria. This method could be implemented for future designs and could become a bench mark for evaluating new systems in new or existing cockpits. It will also allow the applicant to vary software, hardware or procedures in order to achieve approval. This method should be reviewed when discussing the new harmonised FAR/JAR 25 certification requirements for Human Factors.

7 Human Factors of Automation: The Regulator's Challenge

HAZEL COURTENEY

UK Civil Aviation Authority, Great Britain

Introduction

There are two facts that are probably well known to readers of this book. One is that the automation on aircraft flight deck systems, and in other areas of aviation (such as ATC), has grown enormously over the past few decades. The other is that around three quarters of accidents in aviation are now attributed to human error. Taken together, these suggest that those responsible for safety might beneficially examine the relationship between human error and developments in automation. They may do this to ensure that automation is not actively contributing to the potential for error, and also to seek any means by which it could be used to reduce the likelihood, or adverse consequences, of those errors that do occur.

A system, in the true sense of the word, consists of all of the elements that are required to conduct the task at hand. In aviation, most of the functions require, at some point, a human element in the system for the task to be carried out. Civil aircraft do not, as yet, fly unmanned; nor do they maintain themselves. Therefore the traditional approach to airworthiness, which takes little account of the human element, is incomplete in important ways.

The organisations that are responsible for setting and enforcing minimum standards in all areas of civil aviation (including aircraft design) are the national aviation authorities, such as the UK Civil Aviation Authority (CAA), the US Federal Aviation Administration (FAA), and their national equivalents elsewhere. In Europe, many of the national aviation authorities have formed an association known as the Joint Aviation Authorities (JAA). Members of JAA have agreed to co-operate in developing and implementing common safety regulatory standards and procedures. This will hopefully make it much simpler for aircraft manufacturers to work with aviation authorities in Europe.

There are agreed requirements for aircraft design, and these set the boundaries on the minimum standard that can be issued with a Type Certificate, (which is needed to operate the aircraft). Within the JAA, these requirements are known as Joint Aviation Requirements (JARs). The design requirements for large fixed wing aircraft, for example, are known as JAR-25. These requirements have served the technical aspects of aircraft design very well, and their success is evident in the remarkably low rate of serious accidents now arising from technical causes. The issue of human error, and its relationship to design, has been less clearly addressed within the aircraft design requirements.

There are probably several reasons for this. It is a relatively recent viewpoint that certain design features actually encourage error to an unacceptable degree. Some people still believe that all errors can be eradicated by training, or that if a person makes an error it is entirely their own fault. Many users may have been able to operate a certain design correctly before one makes an error, so it is not easy to 'prove' the design feature to be unsafe. It is not a simple matter to describe the aspects of design that might be problematic in a generic manner that would be applicable to all future possible designs, and set the 'minimum acceptable' standard in clear and objective terms. This makes it difficult to set the design standards.

The oversight of military aviation is somewhat different, but the issues are comparable. Within technical development programs, there is a clear benefit to defining the requirements for acceptability of a design product in terms of its relationship to human users, because this is likely to determine its ultimate success or failure as an application. There is a need in both civil and military systems to explore the relationship between automation design and human error, and learn how to ensure that new automation design is acceptably safe.

Human error falls into several areas of responsibility, such that it is rarely - if ever - possible to say with any certainty that the reason for an error was definitively the design, without some relevance being given to selection and training, commercial pressures, airline operating procedures, interactions with external parties and systems, or failures in individual or team performance. The human pilot not only interfaces with these various parts of the aviation system himself, but actually forms the interface that connects these areas to each other, through him. For example, the flight deck design only interfaces with the operating procedures through the pilot; the flight training program can only interface with ATC procedures through him (Figure 1).

Figure 1: The pilot forms the interface between different areas

If different parts of the system are not entirely compatible, this interface becomes strained, and may fail (Figure 2). This failure is known as 'pilot error' or, more recently, a 'human factors' issue. In general, the individual parts of the aviation system receive a great deal of attention, in terms of both management and regulation. The interfaces between them receive rather less, because they are difficult to quantify and have no clear ownership. Achievement of highly compatible interfaces may require individual areas to expand the boundaries of their task beyond its current definition. Ensuring the adequacy of the interface to the pilot, and through him, the interface to other areas, may not necessarily be defined as a program task in civil aircraft development. Cost recovery of such activities may be deferred and indirect.

Business culture strongly favours meeting requirements as defined, and does not tolerate the costs of exceeding that task. Current trends are probably increasing this tendency, with separation of activities into cost centres, individual accountability against key measures, and direct recovery of costs. This could mean that business culture increasingly penalises the activities that are needed to address human factors issues in the aviation industry.

Most companies - and individuals - will behave in the way that is most rewarding for them and their shareholders. Yet, it is not unknown for any industry to systematically penalise the very activities that it seeks to encourage. This is known as '*dis-integration*' (Courteney, 1996), the tendency for the system of rewards to actively impede the progress that the greater good requires. This may be the case in achievement of good 'human factors' design in some areas of civil aerospace. It is probably not realistic to hope that the general business culture will spontaneously change in order to address the human factors issues. Therefore a different solution is needed.

Figure 2: Simultaneously demanding areas stretch the pilot

What action is being taken?

The aviation authorities in the western world have not ignored these issues. In 1994 the UK CAA decided to recruit a human factors specialist (the author) to draft some new 'human factors' requirements for flight deck certification, (in co-operation with other experienced personnel). The criteria chosen to form the basic 'requirements' were largely drawn from the published literature, and the research that had been conducted in the area, which criticised some flight deck design, (and the Flight Management System (FMS) in particular) for not incorporating good human factors considerations into the design, and for being difficult to use, vulnerable to crew error, creating high workload peaks, and presenting inadequate feedback (for example, the FAA Human Factors team report, 1996; Sarter & Woods, 1994a).

These draft requirements then progressed to the JAA Human Factors Steering Group (HFStG), who created a dedicated subgroup to consider and develop the proposal. In 1998 the subgroup gained formal endorsement from the JAA HFStG for the version of the proposed requirements and supplementary material that is presented below. This has been offered as a position for the consideration of a formal 'Harmonisation Working Group' (HWG) between industry and authorities from Europe and the US, scheduled to begin early in 1999. The initial proposal has been offered for Joint Aviation Requirements JAR-25, the design requirements for large aeroplanes. It is intended that, in the future, a similar requirement would be developed for small aeroplanes and helicopters. Supporting requirements (shown below) have been proposed for JAR-21 (Requirement for Certification Procedures), referring to the process by which the product is developed.

Additional Regulatory Material proposed for JAR-25 as being more appropriate to product design:

a)   The design of the integrated Flight Deck Interface must adequately address the foreseeable performance, capability and limitations of the Flight Crew. (See ACJ-25.xxx(a)).

b)   More specifically the Authority must be satisfied with the following aspects of the Flight Deck Interface design. (See ACJ-25.xxx(b)).
1)   ease of operation including automation.
2)   the effects of crew errors in managing the aircraft systems, including the potential for error, the possible severity of the consequences, and the provision for recognition and recovery from error.
3)   task sharing and distribution of workload between crew members during normal and abnormal operation.
4)   the adequacy of feedback, including clear and unambiguous:
i)   presentation of information
ii)   representation of system condition by display of system status
iii) indication of failure cases, including aircraft status
iv)   indication when crew input is not accepted or followed by the system
v)   indication of prolonged or severe compensatory action by a system when such action could adversely affect aircraft safety.

Explanatory Material proposed for the expansion of JAR-25 requirements:

a) The requirement applies not only to new flight deck designs, but also to installation of supplementary equipment, and modifications. Where supplementary or modified features are presented, their assessment should not be limited to those features in isolation, but give due regard to their use as part of the integrated flight deck.

b) In order to demonstrate compliance with the requirement the manufacturers will need to justify decisions made. This should be achieved by the presentation of a clear rationale and, where appropriate, objective evidence, for design decisions with implications concerning the safe conduct of the flight by the crew. The Authority shall monitor the development process and be invited to attend evaluation trials. Acceptance of the product will be subject to assessment of the final completed design by the Authority.

The certification of the aircraft design will be conducted by assessing the product. It is the product design that receives the approval and the Type Certificate. However, it is difficult (if not impossible) for a certification team to examine every permutation and future eventuality when assessing a product design. Therefore, it is important that the manufacturer presents evidence that they have met the JAR-25 requirements through thorough development practices. This evidence would be a 'means of compliance', that is, a way in which their claim to have addressed, say, the 'effects of error' can be supported. Some possible future 'means of compliance' are discussed later in this chapter.

Requirement for certification procedures

Good 'human factors' qualities in flight deck design depends upon properly planned and conducted activities during the design process. This has led to a new requirement for JAR-21, the Requirement for Certification Procedures. The proposal (shown below) requires that the manufacturer (the 'applicant') submit a plan showing how human factors issues will be addressed within the aircraft design and development program. This would ensure that the subject received due consideration from the outset, and that good intentions were not discarded during the program due to inadequate resources or diverted funds. It is important to be clear that this is not a 'certification' of the process. A documented plan would be required as part of the Certification Procedure, and the certification of the *product* would include scrutiny of the evidence - probably generated from the development process - that the product design criteria had been met. The evidence would be used to demonstrate the compliance of the product design. The process *itself* is not the subject of the certification, but receives approval as a planned way to generate evidence to support the application for certification of a product. One reason for this is that it is always possible to 'go through the motions' and pay 'lip service' to a process, and hence produce an unsafe design. This cannot be allowed to pass into service simply because it resulted from a 'correct' process description (example, ISO 9001 is a 'process based' standard, but does not guarantee the merits of the product).

Additional Regulatory Material proposed for JAR-21 (JAA Requirement for Certification Procedures) as being more appropriate to process approval:

(a) The applicant shall address JAR-25.XXX(b) with documented proposals for approval by the Authority prior to any compliance demonstrations or evaluations. An acceptable plan will include:

1. The functional philosophy for the flight deck design and any integral automation.
2. The means of evaluation to be used by the applicant.
3. The development schedule, including evaluations and reviews where the Authority is participating.

(b) The plan may be amended, if justified, documented and agreed by the Authority. Adherence to the plan will be monitored.

To avoid hindrance to the manufacturers development program, a policy statement is required regarding availability of Authority personnel.

Explanatory Material for the expansion of JAR-21 requirements:

(a) The applicant will define the means by which the Flight Deck Interface design, and changes to the design, are authorised within their organisation.
(b) The functional philosophy for flight deck design and any integrated automation will be used and applied by all relevant parts of the organisation, including sub contractors.
(c) The design must be compatible with the pilots abilities and limitations (JAR-25.XXX). This can best be achieved through representations of the system receiving timely evaluation by approved test pilots, and the structured involvement of operational users (e.g. performance on part task simulation) throughout system development. Features that are novel in concept or in presentation should receive particular attention.
(d) Guidance for flight deck development practices is contained in Advisory Material - Joint (AMJ-21.XXX). However, the documented plan should cover the following topics as a minimum:

♦ The scope and organisation of the resources required for design and evaluation of human performance and limitations aspects of the flight deck.

♦ The development schedule, including milestones and reviews.

♦ The functional philosophy for flight deck design and any integral automation.

♦ The means of evaluation, including details of subject flight crew proposed for structured trials and the levels of simulation to be employed.

♦ The means of systematic evaluation of the effects of flight crew error.

♦ Details of configuration control methods.

♦ Details of supporting information to be used, including research data and in-service reports.

♦ The methods for evaluation of training effectiveness, under normal and abnormal circumstances.

♦ The means of evaluating operational suitability, including integration with current or proposed navigation and communication systems.

A model for future development

The JAR-25 proposal above offers high level human factors criteria for the judgement of the certification team, and the associated JAR-21 material endeavours to ensure that the manufacturer addresses these issues in a systematic and competent manner. However, if this material, or similar, progresses through the HWG process, there may be a benefit in augmenting it with supporting advisory material. This would aim to offer manufacturers some guidance on 'means of compliance'. That is, what can they do to convince the authorities that they have achieved these criteria? What follows here is not a formal view from the authorities, but a personal view from the author suggesting how this area might develop.

The basic proposed high level criteria for JAR-25 have already been presented above, but in summary:

Figure 3: Human factors principles proposed for requirements

These are all sound human factors design principles, but they are probably not sufficiently detailed to be able to clearly determine whether they have been met or not. Remember that in regulation, the question is not 'what is best' or 'what should we aim for', but 'has the design achieved a *minimum acceptable* standard'. This is often a difficult concept for those in design or research, who instinctively want to say 'but it would be better if the design had so and so'. Their solution may well be a better one, but that is not the question being asked. The question is 'is it acceptable in the form presented?', and the requirements should, ideally, allow that discrimination to be transparent.

One way to address this within the design process might be as follows. For any planned product, the design documentation would begin by stating its own specific design requirements that aim to meet the general requirements. Take 'effects of error' as an example. What specific product requirements could be used? Much will depend upon the system, but a different requirement style might be useful at this product focused level. For example, it may be decided that appropriate philosophy for addressing the 'effects of error' in a flight deck system would include statements such as:

♦ 'a single slip or lapse (e.g. incorrect data entry) should not be capable of progress to a potentially hazardous outcome without direct and compelling feedback to the crew'

♦ 'no single crew action with a hazardous consequence (e.g. in an aircraft, closing down the only remaining propulsion, or all hydraulic power) should be possible without a direct challenge from the system that must be positively overridden by the crew. Such challenges should not be presented when the action is not hazardous, to prevent it from becoming perceived as routine.'

These statements are not offered as a recommended philosophy, because, for example, the technology may be such that the second item is not realistically achievable without introducing other problems of reliability. These are simply offered to illustrate a style that may allow the design philosophy to be evaluated early in the development process, and give some definite, objective pass / fail criteria, whilst permitting freedom for the design solution to be developed. They would also ensure that basic human performance considerations were incorporated into design decisions at an early stage, not just as advice or good practice, but as requirements.

Demonstrating compliance

Following this level of requirements, would be the means of validation, that is, how would compliance with the requirement be demonstrated?. This is already mentioned in the proposed JAR-21 material, but it seems useful to offer manufacturers some validation methods that are likely to produce an acceptable result. To demonstrate 'ease of use', this would almost certainly involve the iterative user trials, but continuing with the 'effects of error' example, a more systematic method needs to be employed. It is not at all satisfactory to simply record that during user trials, no-one made a certain error, and conclude from this that therefore the error cannot and will not occur. In certification, the permissible failure rates being considered are very, very small, 10^{-9} for failures with potentially catastrophic consequences. If the potential errors are considered to have comparable rates of occurrence, then it would be a very long user trial that would identify them all. A better method is needed.

There are many who would want to employ quantitative requirements here (see the chapter by Singer elsewhere in this volume). Such requirements would utilise error rates that have been obtained by expert judgement or laboratory experiment on what they believe to be generically comparable tasks. This is not, I believe, a good method to evaluate the details of new proposed designs because, as suggested above, the reliability numbers can be somewhat unreliable. This is not because the manner in which they are determined is poor, but simply that they can be changed so much when there is a change in any feature of the design, the task details or circumstances, the person under test, the means of measurement, or any combination of such changes that can interact in unpredictable ways. The use of data gained with similar systems in service may give broad indicators if the information is reliable, which it probably is for large events (how often do airliners stall?) but much less so for small events that might be considered (in isolation) to be relatively trivial. These may not necessarily be reported, (as suggested by the study results reported later in this chapter).

At the present time, the regulatory advice for System Design and Analysis (AMJ-25.1309 para 8.g(5)) states that "quantitative assessments of the probabilities of crew error are not considered feasible" and also that "Probability levels which are related to Catastrophic Failure Conditions should not be assessed only on a numerical basis, unless this basis can be substantiated beyond reasonable doubt" (para 8.d(1)). Substantiation beyond reasonable doubt seems to me to be unlikely. In 1991, the Advisory Committee on the Safety of Nuclear Installations Study Group on Human Factors, led by Dr. D. E. Broadbent, issued a report titled 'Human Reliability Assessment - a critical overview'. Their conclusions were that Human Reliability Assessment (HRA) was "still in its infancy", that it yielded "useful but imprecise results" that must be used "intelligently and with discretion", and agreed that whilst 'skill based' errors (slips and lapses) were amenable to some degree of prediction, 'rule based' and 'knowledge based' 'mistakes' are much more difficult to address. In their Executive Summary (item 12), they comment that:

"Factors that may reduce the accuracy of HRA include systematic errors of human judgement on the part of expert assessors, lack of generality in the evidence derived from laboratory experiments, and paucity of field data derived from records of past operations. It is disturbing that few studies have been made of the validity of methods, that is, the extent to which the assessed probability agrees with a known true probability, or with the results of some other method of HRA. When validity studies have been performed, the results are not encouraging."

Reason (1990, 230-231) notes that expert estimates or 'Absolute Probability Judgements' are as good as any method for generating error probabilities but, quoting Williams (1985) that "The developers of Human Reliability Assessment techniques have yet to demonstrate, in any comprehensive fashion, that their methods possess much conceptual, let alone empirical, validity". Returning to the subject of 'mistakes' (which are not simple 'slips and lapses' in carrying out the task, but have to do with actually trying to do the wrong thing) Reason (p.249) notes that 'it (is) clear that, current probabilistic techniques (are) inadequate to capture the likelihood of operators basing their recovery actions upon an incorrect mental model".

HRA methods are probably useful at the conceptual planning stage of overall system design to get a general picture of how human performance is involved in producing a safe system, but not at the design approval or certification stage where it is necessary to discriminate between design details and it is important to ensure that specific risks are identified.

Another means of validation for addressing the 'effects of error' is currently under development by the author. I will refer to it as Human Hazard Assessment (HHA). This does not use error probabilities, but assumes that anything that can go wrong will, eventually, go wrong, and as such must be considered as a real possibility. The general approach is derived from the engineering Functional Hazard Analysis (FHA) and Failure Modes and Effects Analysis (FMEA) techniques, but instead of the start point being a technical failure, it begins with the crew task. It mimics their qualitative but systematic consideration of failure through a tabular format, but progresses to ask, not only what is the 'Crew Action Required' as does FHA, but also, for example:

♦ What signal or situation prompts the crew to take the action (a new ATC instruction, a visual assessment of the approach path, a loud noise like an engine failure)?

♦ Is this the only possible interpretation of that signal, and if not, what other causes or interpretations are there? (e.g., the loud noise might alternatively have been a burst tyre)? (Would the required crew action then be different?)

♦ What latent conditions could precede the action (previous errors - such as ATC instruction misheard - or technical failures)?

♦ What other actions could the crew do instead (e.g., they could do nothing, the correct action but delayed, the reverse action, the right action on the wrong control, or select the wrong item, wrong digits, wrong mode etc)?

♦ What would be the direct / immediate consequences of that incorrect action?

♦ What immediate indications or feedback would there be?

♦ What other indications or feedback would be available (subsequent system behaviour, secondary sources, cross checks and comparisons)?

♦ What other secondary defences exist (special training, crew procedures, checklists, external monitoring e.g. ATC)?

♦ What could the eventual direct consequences be?

♦ Could this error create a latent consequence, e.g., could it make a subsequent error or technical failure more significant in safety consequences?

♦ What could the eventual indirect or combined consequences be?

♦ Should mitigating action be taken by the design or other means?

This method might be a means to validate design aims such as the examples offered above. By progress through the table, it should be possible to identify specific errors that could, in themselves or in combination with a secondary failure, result in a potentially hazardous situation.

The manufacturer might be asked to nominate the proposed method of analysis in their plan and, if the plan were accepted, to show that this had been conducted and that the relevant items had indeed been addressed as the design progressed. (AMJ-25.1309 para. 8. suggests that "Early agreement between the applicant and the Certificating Authority should be reached on the methods of assessment to be used".) Of course, such methods will never be entirely objective or 'water tight', since there would still be some negotiation concerning the conclusions drawn about risks and mitigations. For example, incorrect mental models may be considered the human equivalent of a 'common mode failure' in engineering terms, and form an important element in risk assessment.

Promises and pitfalls of user involvement

User involvement is the subject of much misunderstanding and debate. It is all too easy to conduct user trials that have high surface validity and look convincing but, in reality, are at best worthless and at worst actively misleading. This topic is covered at length in Courteney (1998a), but it is worth a review of some of the major issues. In civil aviation, the first stumbling block is who the 'representative users' should be. This is much easier for military designers, who have more direct access to operational pilots and, typically, know which nations are likely to buy the aircraft. In the civil market, crew attendance costs can be prohibitive, but there are also issues of commercial confidentiality and the awareness that almost any country (and thus culture) may buy and fly the aircraft. In addition, the customer airlines will be far more at liberty to select who is to represent their pilots, and are likely to choose technical or management pilots, or at least their most competent line pilots, to represent the company. Therefore, acquiring a truly 'representative' cross section of future users becomes much more of a challenge, and may have to be achieved by requesting every 'nth' pilot on the crew list from a certain fleet, despite the inconvenience and possible annoyance to the customer. No manufacturer can be expected to enlist and accommodate users from every culture that might ultimately buy their aircraft.

Then, there is the issue of how opinion is measured, what is tested and how, and what the results really mean. The fact that a design is successfully used and liked by a sample of n pilots, and none of them made any errors, does not prove that it is acceptable. The fact that crew say they can easily understand what is going on does not prove that they actually do correctly understand, hence the concept of 'miscalibration' described by Sarter & Woods (1994a). Crews were often found to claim - and believe - that they understood the automation when deeper exploration demonstrates that they did not.

As a rule of thumb, line pilot assessments should be used for the following reasons:

♦ to calibrate the use of the design with a representative interface to the rest of the operational world, one with a similar level of ability, training history, daily routine, and experience of day to day procedure, as the final users
♦ to help spot task disjoints, risks of error, misunderstanding or incorrect assumptions; incompatibilites with the wider system or clashes with existing conventions, that may not have been identified by other means
♦ to help maximise ease of use and identify any specific difficulties or ambiguities

In general, operational pilots objections should all be at least considered but, to a large degree, their endorsement should be ignored. That does not mean that their opinion is not of value, but the fact that someone who is not trained in evaluation successfully uses a design or cannot see any problem with it does not prove that there is no problem. However, the fact that they *do* make errors or *can* see a problem should be taken much more seriously, and explored. For making a decision about the adequacy of a design, (once the identified risks have been addressed as far as possible), trained test pilots are needed. In my own experience, using operational line pilots to run through task scenarios can identify new risks and potential misunderstandings that were never dreamed of by the design team, test pilots included. Real users can misuse the system in ways that are highly original and mysterious. For these reasons, the proposed requirements above include specific reference to *the structured involvement of operational users* (within the JAR-21 proposal).

The difficult part, from the regulators viewpoint, is in providing an *a priori* description of what, in terms of user involvement and interpretation of the results, is to be considered adequate. There are so many possible permutations of circumstances and methods that a set of 'rules' to determine what will be regarded as acceptable is nearly impossible to produce. This means that the authority needs some means to gain confidence that the user involvement will be conducted properly, and will not be reduced to sitting operational pilots in front of prototypes and asking them to 'comment'.

Organisational approval and capability maturity models

A design or manufacturing company needs a basic level of education and competence in order to conduct the planning and validation in a meaningful way. This competence varies tremendously between organisations, from the 'cutting edge' leaders in their field, to the other end of the spectrum where a breathtaking naiveté and lack of awareness can still prevail. It may be, in the future, that human factors capability becomes another area that must demonstrate a competence to at least an acceptable minimum level, in order that the company approval is granted. Because the specialisation is relatively new and standards are so variable, one possible approach would be to use the concept of Capability Maturity Models (CMM) that have proved so popular in the software industry. This notion is also explored and developed in Courteney & Earthy (1997).

Table 1: Capability maturity models for human factors?

Stage	Typical Attitude
Ignorance	"We don't have problems with human factors"
Uncertainty	"We don't know why we have problems with human factors"
Awakening	"Do we always have to have these problems with human factors?"
Enlightenment	"Through management commitment and improvement of human-centred processes we are identifying and resolving our problems"
Wisdom	"Human factors defect prevention is a routine part of our operation"

By using a multi-level 'maturity' index, it would be possible for authorities and procurement agencies alike to continue working with companies whilst acknowledging that some of them need more oversight than others in this respect. A good outcome of such a system would be that only the most basic level of achievement might be acceptable for a manufacturer to continue to be an 'approved' company in the first instance. However, their perceived need for improvement in this respect would be explicit, and hopefully a motivating factor in improving their methods. Companies assessed to have a high level of maturity could use this as a positive selling point to customers and agencies elsewhere.

In addition to professional competence, there is an issue of resources and facilities. Cost and effort need to be planned into the development program, and sometimes complex facilities are needed. The chapter by Singer in this book suggests the extensive use of simulators. It is true that these are an invaluable tool, but for regulators to actually require that manufacturers possess, or have access to, full simulation of their designs, would be a serious step with significant cost implications. To require simulator trials but to exempt those who do not own a simulator would be a nonsense, and possibly discourage the ownership of simulations.

Taking all of these points into account, a possible model for the future might, at this stage, resemble a pyramid, as each successive layer becomes broader and lower level, but supports the focus of the one above:

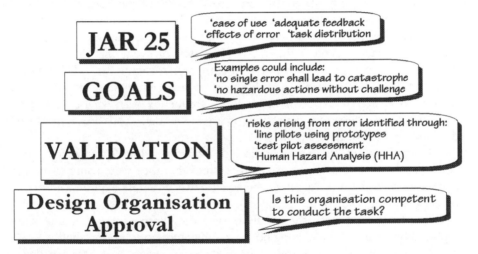

Figure 4: Human factors requirements supported by means of compliance

Flight deck events: the flight management system as a case study

The criticisms of existing regulation have come from various sources, some of which are very knowledgeable about certification practice, some from laboratory or simulator experiments, and some from team investigations based upon expert opinion and dialogue with manufacturers. However, until recently, one source of information appeared not to have been included in the debate: that is, operational data on the events that actually take place on the flight deck during commercial service. Courteney (1998b) reported a case study of events involving the FMS, recorded during 2066 sectors flown in public transport operations. The results made an interesting comparison with the literature predictions and popular criticisms of FMS technology.

In general terms, a little more than half the sectors recorded showed no events connected with FMS, whilst the remainder reported an average of almost two events each. Overall, this meant that the event rate was slightly less than one event per sector. From the current wisdom on criticisms of FMS, it might have been expected that there would be numerous reports that crew did not understand the system logic, made errors that caught them out, suffered from peaks of high workload, inadequate feedback, or lost situation awareness. It is true, such events were reported and, due to the nature of such things, may have been under-reported, but they were a small proportion of the total returns.

The large majority of reports concerned aspects of the FMS that have, in recent debates, received much less attention. The largest category of reports concerned the necessity of crew activities to 'work around' the system features, in order to achieve their tasks. This type of event was closely related to other large categories of reports: 'incompatibility with the ATC system', and 'automation discarded'. In addition, equipment faults and incorrect data (in the system) were reported much more frequently than might have been expected. The following figures summarise the data and are reproduced directly from the original report.

Comments on the study indicated that the pilots usually felt quite confident in their understanding of the system that they were using (although some indicated that this might not be equally true with less experienced pilots, or crew serving with other operators). The notorious 'what's it doing now?' question from pilots seems gradually to be fading with experience, and being replaced by 'look what it's doing now, we'll have to work around that!'

Comparing the relative frequency of different event types between fleets, the most striking contrast is between relatively new aircraft (such as the B777 and A340) and those with a well-established service record (A320, B757, B767, B737). The new fleets reported more cases of 'system logic not understood' and fewer 'workarounds', suggesting that, with fleet experience, pilots learned to figure out what the system was actually doing, and how to work round it. This is probably not how the designers had intended their systems to develop.

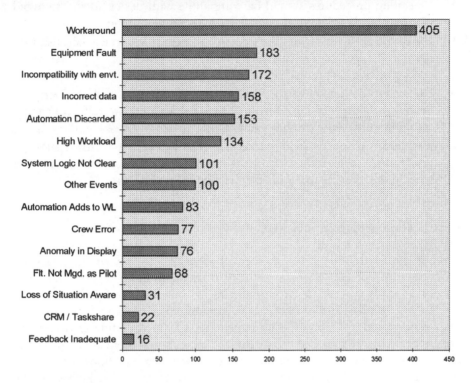

Figure 5: Results from FMS line study

Whilst under-reporting is always a possibility, it was surprising to find that *none* of the events recorded during this study had been reported by any other means. It is true that many of the events might not have been worthy of specific reports (such as routine workarounds) but others, such as factually incorrect data in the system, might well have been items that manufacturers and operators alike assume would be highlighted, should they occur. Not so. In general, what these data suggest is that:

♦ Validation of FMS design requirements against user experience and the ATC environment is incomplete.
♦ Validation of the FMS design functionality and data against known requirements is incomplete.
♦ Existing training is not necessarily sufficient to ensure the crew can predict aircraft behaviour at all times.

◆ The system design must assume that crews will make routine errors and accommodate this characteristic.

◆ Operating procedures should be compatible with design intent, to avoid problems such as late (and unachievable) reprogramming.

◆ Flaws and difficulties with the system can persist in service, because they are not always - or even usually - dealt with effectively (by a reporting and corrective action mechanism).

◆ Safety implications may arise indirectly from aspects such as increased head down time, erosion of manual flying skills, and late runway changes resulting in rushed reprogramming.

◆ Suppliers cannot be held entirely to blame because many issues go unreported; crews cannot be held entirely to blame for not reporting when they often see no result for their reporting efforts.

Certification proposals revisited

Having gathered this data, it seemed sensible to revisit the existing certification proposals. There is nothing in this data that suggests that these proposals are inappropriate or wrong, simply that they are incomplete. The four main criteria, and systematic validation coupled with iterative user involvement that will bring greater experience of the operating environment, are confirmed to be important. However, it appears that there are many issues arising from the system working in service, where its interactions with the air traffic system and the operational world are not always as seamless as one would hope. The majority of these events are not, in themselves, a direct threat to safety, but even the minor items add load and distraction. These may be repeated many, many times, before they occur in combination with other factors that lead to a safety risk. In the cases of 'incorrect data', they may lead directly to a safety risk if combined with the crew, on that occasion, not spotting the anomaly.

This suggests two possibilities. One option would be that the system must be so well validated before it enters service that these flaws are reduced to a negligible level, (a task so difficult and time consuming that one wonders whether it is realistically feasible). A higher standard of software integrity could potentially be enforced. Currently FMS software is required to be produced to DO 178 level C and this could feasibly be raised to level B; indeed, some suppliers are already adopting this level voluntarily. However, from the data collected it would seem that this would not necessarily solve the problem of workarounds or incompatibilities with the reality of ATC. Taking a pessimistic view, it could simply result in more highly reliable incompatibilities. Alternatively, a system could enter service at the current standard, but with a much improved method for collecting and acting upon in-service reports. This would suggest that, at the point of approval or certification, the applicant (manufacturer) would have to demonstrate that a mechanism was provided to collect the appropriate data, and that resources were already earmarked to deal with the ensuing corrective work. A possible side effect of such an approach might be to encourage watchfulness and reduce complacency on the part of the crew, because it would emphasise the fallibility of technology and promote the need to challenge and cross check. The one major danger, here, would be the possibility that manufacturers would begin to rely upon this post implementation corrective period to allow standards of integrity in the initial design to drop, promising to 'fix it later', but meeting their delivery dates effortlessly. This would require vigilance on the part of the regulators to ensure that the existing standard for product certification was maintained.

Another issue raised by this study is the relationship between the design, the operational procedures, and crew training As suggested in the Introduction, it seems there is a gap between the responsibilities of 'design' and 'operations', both in the industry, and within the regulatory authorities. A visit to one FMS supplier organisation highlighted the difference between the way in which they have designed the system to be used, and the way the flight crew expect to use it. Crew reports that the FMS did not perform well in certain situations were met by engineers cries of 'Good grief, they're not trying to do *that* with it are they?! It was never designed to do that!' In a similar vein, when a system is assessed for certification, there is no route for the certification team to decide that this design may be used safely if, and only if, special training or specific operating procedures accompany it into service. Perhaps there is a need for a design certification team to have a mechanism to raise 'special training items' (STIs) and 'special procedural items' (SPIs) such that the more challenging items in the design can be integrated with the operational system to best effect. In fact, to restrict the use of STIs and SPIs to the certification team could be a missed opportunity. This mechanism might be used to even greater effect as issues arise within the manufacturer, during the validation program. One means to use the STIs would be to consider them for addition to the Type Rating syllabus; SPIs might be passed to the Flight Manual (such as emergency checklists) or to company operating procedures. As with the post implementation reporting, the difficulty here would be in resisting the temptation to abuse the system, by simply making every marginal compliance a STI or SPI, instead of making the proper design adjustments. Manufacturers would probably exercise restraint in their use because they would want to minimise the operational burden that accompanied the aeroplane, but they could pose a temptation for certification teams.

Finally, there is a need for the whole design culture to learn more from the in-service experience of their products. This is not a trivial matter, because the huge number of hours flown on popular aircraft world wide makes for a vast quantity of experience to record, report and analyse. Worse, the liability laws are somewhat unhelpful in this respect, because once a manufacturer has information about a certain feature that could jeopardise safety, it will have seriously adverse consequences for him if an accident occurs involving that feature, before he has addressed it. If, on the other hand, he has declined to collect any such data, he will not be held so responsible for an accident because he did not have any information to indicate that there was ever a problem. A parallel situation exists within the operators, who strongly resist identifying 'hot spots' in crew performance problems for their training programs, because this would mean admitting that they are aware that a problem exists, so if there was an accident, their liability could be raised. These are yet more examples of a *'dis-integrating'* environment - there are very real punishments for trying to do the right thing!

From these observations, a more balanced model could be formulated, taking the original model and adding an 'inverted pyramid' mirror image requirement that comes into effect post implementation, making the model into the shape of an 'hourglass' instead of a pyramid.

This is one persons view, and does not represent any official plans for the future. However, similar concepts have proved successful in the world of software, where the evaluation of acceptability can be equally difficult to pin down. The challenge of drawing that elusive 'line in the sand' that separates 'shades of grey' between acceptable and unacceptable can never be other than by human judgement in all but the sim plest of cases, because the factors are many and varied. Such is the case in all complex evaluations, whether they be in aircraft certification, medicine, or law. To attempt to tie such issues down to numerically measured quantities is to negate that most valuable human ability to consider such multi-faceted issues better than any mechanistic approach. Still, there is much to be gained from ensuring that a certain level of risk reduction has been systematically conducted, and that a further layer of feedback and correction will accompany the aircraft into service. In addition, this model represents a first step in formally linking design certification to operational issues such as procedures and training. This 'linking' may be another area that could benefit from expansion, ensuring that the design and the operational environment are compatible and do not depend upon unrealistic expectations of human performance.

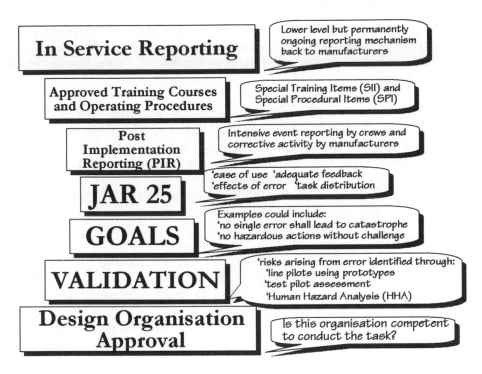

Figure 6: The hourglass model of regulation

Conclusions

From the chapter by Singer and the topics discussed above, it seems clear that there is a need to improve the design requirements and overall regulatory approach to flight deck systems. The means by which this will be achieved is still open to debate, although there is progress being made, as outlined in this chapter. Regulatory action on the human factors aspects of flight deck automation has been driven by six basic factors:

1. Air traffic is increasing. If the number of air accidents is to be kept down to a tolerable level, the accident rate per flight must be actively reduced.
2. Crew error on the flight deck is a causal factor in the majority of accidents; technical causes are now less usual.
3. Automation is the aspect of flight deck design that is changing most rapidly, and can be relevant to serious crew errors.
4. The existing design requirements for certification of flight deck design were written before technology had progressed to current levels, and are not strong in human factors aspects.

5. There has been criticism of flight deck design from the human factors community, and this reflects on manufacturers, operators and regulators.
6. Discussion and interest in the industry has gained momentum up to a critical mass.

This year, for the first time, the JAA has requested that a design certification team should include a human factors specialist. No doubt this project will generate many more considerations that have yet to be discovered, and ideas for the 'best approach' will change and grow. The existing proposals may or may not survive this scrutiny, but whilst we cannot be sure exactly what is going to happen we can, at last, feel confident that something will.

Acknowledgements

Thanks to British Midland for their assistance with development of Human Hazard Analysis for maintenance tasks; to Smiths Industries for their assistance with development of Human Hazard Analysis for Flight Management Systems, and to the Air Traffic Management Development Centre for their assistance with development of Human Hazard Analysis for new air traffic systems.

8 Extracting Data from the Future – Assessment and Certification of Envisioned Systems

SIDNEY DEKKER AND DAVID WOODS*

Centre for Human Factors in Aviation, Linköping Institute of Technology, Sweden
*The Ohio State University, USA

Introduction

How do we gain empirical access to systems that haven't been built or put together yet? In other words, how can we get human performance data from environments that do not yet exist? This question, in short, forms the envisioned world problem - the problem that confronts developers of large-scale human-machine systems when they wish to assess the impact of those systems on human performance (Dekker, 1996). The envisioned worlds that elude researchers and developers alike tend to profoundly shift roles of humans (e.g. in the future, we will have air traffic management by exception; we will conduct coronary surgery with remotely controlled robots) while otherwise staying rather ill-defined. They also emphasise pushing the technological frontier (e.g. getting various anti-collision and remote sensing technologies to work in harmony), while being unable to sort out how such technologies will alter the future system's interactive work, its error potential or vulnerability to breakdown.

Since the envisioned world doesn't exist yet, and there are no practitioners who work in the future, what should human factors assessment - let alone certification - be based on? Lacking empirical access, envisioned world problems are often resolved through the substitution myth. People readily adopt the idea that human work can be replaced by machine activity without creating any reverberations for the larger human-machine ensemble and the kind of work and interaction that goes on within it.

One such envisioned world is future air traffic control. In air traffic control there is much pressure to automate routine portions of the human task, putting controllers at a larger supervisory distance. This will allow higher system throughput and consequently allow air traffic centres to absorb ever-increasing traffic loads. No one knows how controllers will perform in the role of distant supervisor, in the role of exception manager. But because of the lack of empirical data to the contrary, assumptions of human-machine substitutability are routinely made here: the machine only takes over routine portions of the task, freeing up the human for more supervisory work, and for the rest nothing is changed. In this chapter, we report on one way to try to crack the envisioned world problem, in this case in air traffic control. Through various future incidents that mimicked the rules and roles of the envisioned system as closely as possible, we were able to generate one kind of empirical data about human performance in the future.

One envisioned world problem: automation in air traffic control

Throughout the world, air traffic control (ATC) centres must be prepared to absorb ever higher traffic loads (NRC, 1998). One vaunted key to accommodating growing throughput pressures is the automation of certain human controller tasks. Indeed, it is widely claimed that a prerequisite for increasing airspace capacity is that controllers step away from controlling every aircraft individually. The basic thought is that while the sky is huge and more aeroplanes will fit in, more aeroplanes cannot be in one controller's head at the same time. There are thought to be hard limits on the number of aircraft that any controller can control at any one time (see e.g. Baiada, 1995). The putative solution is to put more aeroplanes in the sky, but fewer in the controller's head. Automation of parts of the controller's work is then of course necessary. "To permit unrestricted ATC growth, we should first determine how to eliminate one-to-one coupling between a proactive sector controller and every aircraft in flight. The basic requirement is to minimise human control involvement in routine events, freeing controllers to concentrate on the key areas where human skills have most to offer: traffic management, system safety assurance, and dealing with the exceptional occurrence" (Whicher, as cited by Cooper, 1994).

Indeed, the idea of controllers as supervisors at a larger distance has become entrenched as a stereotypical industry solution to traffic growth problems. As example from one aviation magazine: "Controllers - renamed traffic managers - would stand by to intervene to resolve conflicts" (McClellan, 1998). In future ATC architectures, automated tools for conflict detection and resolution are supposed to take a greater role in picking up much of the routine work of keeping aeroplanes on their tracks and away from one another. Controller intervention will be necessary only in those cases when automation is not able to resolve potential manoeuvres that might interfere with other aircraft, when traffic density precludes route flexibility, or when flight restrictions are considered necessary for safety (RTCA, 1995; Nordwall, 1995). In other words, the exception manager will have to resolve conflicts when aircraft or the computers themselves are unable to, and take over control when airspace gets too busy or when other critical parameters are exceeded.

Quantitative benefits, but qualitative side-effects

Technological and procedural innovations are often introduced for their putative quantifiable benefits (e.g. less workload, fewer human errors, higher precision, cf. Woods, 1994a) and future ATC architectures are no exception. With automated tools for separation, people think controllers will have *less* workload in terms of routine tasks (they will only manage exceptional situations and ATC facilities might even be able to do with *fewer* controllers), there will be *larger* traffic throughput, and *higher* accuracy in navigation (GPS) and flight path monitoring (ADS-B) (Coyne, 1995; Cotton, 1995; RTCA, 1995; Baiada, 1995; RMB, 1996; Scardina, Simpson & Ball, 1996; Leslie, 1996).

   All of these promises are of course rooted in the substitution myth - the idea that automation simply replaces human work without creating any larger reverberations for the human-machine ensemble and the sort of work that either does in that ensemble. Automation doesn't replace human work. It changes it. The supposed quantifiable benefits of automation are often offset by qualitative side effects. Previous research has documented that profound changes occur in human roles and human work when humans become more distant supervisors of their monitored process (Hollnagel & Woods, 1983; Hollnagel, 1992; Stix, 1991; Hughes, 1992; Woods & Sarter, 1998). Indeed, experiences from other application worlds suggest that different patterns of function allocation have wide reverberations for the entire human-system ensemble and its collaboration (Hollnagel & Woods, 1983; Dekker & Wright, 1997). Turning the human into a higher-level supervisor often does not reduce human task demands, but changes them in nature. The kind of workload and its distribution over time may change too (Wiener & Curry, 1980; Kantowitz & Casper, 1988; Stix, 1991). In addition, some forms of human error and vulnerability may disappear, only to open the doors for new and unanticipated types of human-system breakdowns (Wiener, 1987; 1988; 1989; Woods, Johannesen, Cook & Sarter, 1994).

Automation and management by exception

It is one thing to try to push the technological frontier and think of ways in which automation could potentially do some of the regular controlling work. This is where the majority of investment in new ATC systems goes. It is quite another to try to anticipate what sort of human work will be created as a result of this, and how this shifts the knowledge and skill requirements, error potential and vulnerability to breakdown in a future ATC architecture. This latter issue is underrepresented in developments of future ATC systems, if anything because it is enormously difficult - it is the envisioned world problem. The reason for not investigating the cognitive consequences of management by exception in ATC is that it is hard to investigate the impact of computerisation in a world that doesn't yet exist. Perhaps many would like to know what consequences such automation will have for a controller's tasks and for system safety, but how are they to find out? This is the research problem with future air traffic control: we must have empirical access to a system that hasn't been built or tried yet in its entirety. Only bits and pieces exist (in the form of various technological innovations: TCAS, ADS-B, GPS, etc.). These do not easily allow us to draw valid conclusions for what a controller's work will be in a complete future constellation.

What doesn't help either is that existing literature on management by exception is largely unable to break out of the substitution myth. It is largely underspecified and does not make precise predictions about the cognitive work that humans will have to do and be good at as exception managers. Below we will first walk through some of the management by exception literature. We will see what questions it leaves unanswered and then take those into future incident studies. These studies were conducted in settings that partially simulated future air traffic control conditions (i.e. future system rules and practitioner roles) in order to gain empirical human performance data about a system that does not exist today.

Management by exception in the literature

Existing literature offers little help in anticipating the qualitative side effects of turning the human into a manager of exceptions. The idea of management by exception in human-machine systems was born out of the supervisory control paradigm in the mid-1970's (see e.g. Sheridan, 1976; 1987; 1988; 1992; Bittel, 1964; Edwards & Lees, 1974; Mackintosh, 1978; Umbers, 1979; Sanderson, 1989). In the words of Wiener (1988, p. 456): "The exception principle states that as long as things are going well or according to plan, leave the managers alone. Don't clutter their world with reports, warnings, and messages of normal conditions". The supervisory controller not only monitors, but also has the ability to intervene in the monitored process when conditions demand. There are a number of ways in which such intervention can take place. To describe the intervention options, supervisory control has typically adopted a continuum of increasing human involvement with the details of the process (Sheridan, 1992; and for more coverage see also Stein, 1992). In other words, the human operator can intervene deeply and take much control away from subordinates (e.g. task-interactive computers), or intervene less deeply and leave subordinates relatively free in their control of the monitored process. Thus, a series of levels of supervisory control is created, whereby human and subordinate control over the process becomes symmetrically apportioned as viewed from top to bottom (see Figure 1).

Levels of Supervisory Control

The subordinate:

1. offers no assistance: human supervisor must do it all;

2. offers a complete set of action alternatives, and

3. narrows the selection down to a few, or

4. suggests one, or

5. executes that suggestion if the supervisor approves, or

6. allows the supervisor a restricted time to veto before automatic execution, or

7. executes automatically, then necessarily informs the supervisor, or

8. informs him after execution only if he asks, or

9. informs him after execution if the subordinate decides to.

10. decides everything and acts autonomously, ignoring the supervisor.

Figure 1: A list of levels of supervisory control (after Sheridan, 1976; 1992)

The list of levels indicates the varying degrees of possible supervisor involvement and alludes to the nature of the human task at each of the levels. But that is about it. The list does not represent the cognitive work that might be involved in deciding how to switch from level to level. It leaves unspecified how the human should decide when and whether to intervene or when to back off (Note that in management by exception, this in some sense translates to the question, what is an exception?). Supervisory control theory offers little further guidance (see Moray, Lee & Hiskes, 1994). According to Sheridan (1987, p. 1249), human monitoring and intervention involves activities such as observing displays, looking for signals of abnormal behaviour, making minor adjustments of system parameters when necessary, and deciding when continuation of automatic control would cease to be satisfactory. These activities are all relevant of course, but their description assumes that evidence on developing anomalies may be unambiguous and that the level to which intervention is necessary is clear to the supervisor.

One answer is that "'exceptions' are pre-defined, and lower-level managers or computers flag exceptions, which are routed to the manager" (Wiener, 1988, p. 456). In future ATC architectures, developers have pre-defined the exceptions: they include potentially dangerous aircraft manoeuvres, traffic density, or other conditions that compromise safety (RTCA, 1995). Computers (Human Interactive Systems, or HIS, in supervisory control language) will indeed largely be responsible for the "flagging of exceptions" (Smith *et al.*, 1998). But it is not hard to imagine that circumstances will arise where evidence on developing anomalies is too uncertain and ambiguous for pre-engineered algorithms. In these cases, the air traffic managers themselves will have to recognise an exceptional situation for what it is worth.

Most theories on supervisory control acknowledge that the supervisor's decision to intervene can be expressed as a trade-off between gathering more evidence and intervening in time (Edwards & Lees, 1974; Umbers, 1979; Sheridan, 1976; 1987; 1988; Rouse, 1985; Sanderson, 1989; Wickens, 1992; Woods, 1992; Rasmussen, 1993; Kerstholt, Passenier, Houttuin & Schuffel, 1996). The more evidence is collected that an "exception" is indeed imminent, the more reliable the manager's decision is of what, if anything, is wrong. However, waiting longer may make the situation worse and corrective action more difficult, or even impossible.

That this can be a tricky trade-off was shown by the introduction of automated process control in steel manufacturing. The 1976 Hoogovens report details how supervisors of newly commissioned process control systems in steel plants had little ability to see the autonomous process in action, and the complexity of process activities obscured sources of problems and malfunctions. To supervisors it was often unclear what kind of intervention should take place, and when or whether they should intervene at all. Unsure of how their manual intervention would interfere with automatic anomaly compensation, supervisors frequently left anomalies to escalate.

Establishing the need for intervention and identifying the best ways to intervene in a (partially) autonomous process is difficult - not only theoretically but also practically. For management by exception (in air traffic control, but in all kinds of other applications as well), we need to explore a set of intertwined questions. What evidence shows that the automation needs help; shows the need for human intervention? Is there any trade-off between intervening early and late? Does the human have any choice in how to intervene or is this dependent in part on when in developing circumstances (s)he intervenes?

Investigating envisioned worlds

In the context of a larger NASA project to study future air traffic control architectures (see Smith *et al.*, 1996) we set out to investigate the kinds of cognitive work an air traffic controller would have to do as manager of exceptions. As discussed, the nature of the controller's cognitive tasks and challenges was hard to predict in detail on the basis of existing literature on management by exception and supervisory control. A series of empirical studies was conducted over a period of more than a year, using active air traffic controllers, airline pilots and flight dispatchers in simulated "future incidents". Each probed different aspects of the cognitive work of (and co-ordination between) various practitioners in the future ATC architecture, in order to characterise management by exception and eventually develop models that may capture its underlying psychological mechanisms.

The challenge in generating results about cognitive work in future air traffic control is that such future architectures exist nowhere in the world. The envisioned world represents a radical departure from existing practice - rendering conclusions on the basis of today's world inapplicable and generally making it attractive to wait for further developments, requirements specification and hardware in order to generate valid results about operator problem-solving in the future system.

But waiting for a more developed world suggests that there exists a trade-off exists between getting in early and generating valid results. We decided not to wait for more future ATC specifications, as experience in almost all developments of human-machine systems has taught that leverage for change decreases as time goes on. The actual trade-off exists between generating valid results that can provide useful input to system development early on and generating valid results that cannot because they come too late. This shifts the dominant source of validity in research on envisioned worlds as compared to more developed or existing operational environments. In the latter, face validity of a simulation tool or experimental set-up is often thought to provide much of the requisite mapping between test situation and target world. In contrast, in research on envisioned worlds, validity derives from (1) the extent to which problems-to-be-solved in the test situation represent the vulnerabilities and challenges that exist in the target world, and (2) the extent to which real problem-solving expertise is brought to bear by the study participants (see Orasanu & Connoly, 1993; Klein, 1993; Woods, 1993). Our studies rated high on both of these measures by (1) creating future incidents, and (2) involving real practitioners who had been prepared for their future roles. In other words, these studies examined real practitioners caught up in solving real domain problems.

Future incidents

Each of the studies was constructed around a "future incident". These incidents contained instances of critical events that could happen in air traffic control systems of any vintage, because of their technology- or time-independent nature. For example, we introduced communication system failures, clear air turbulence, frontal thunderstorms, a cabin depressurisation and a priority air to air refuelling request into the future world to probe the problem solving activities of its exception managers. This method was not intended to prove in general that a future architecture could work in one way or another (for this has been delivered in many underspecified ways, see e.g. RMB, 1996), but rather to explore ways in which it could break down. We elected to use the potential vulnerabilities that a future world could be exposed to as experimental probes, and investigate the cognitive demands on supervision and co-ordination that would have to be met in order to handle and contain them (indeed the envisioned role of the exception manager) (see Woods & Hollnagel, 1987; Woods & Sarter, 1993). Identifying and specifying sources of vulnerability was done in close collaboration with many different domain experts (air traffic controllers, pilots, airline safety officers, dispatchers) from different areas and countries. These practitioners did not participate in subsequent studies.

We invited active air traffic controllers, flight dispatchers and pilots to participate, representing the different user perspectives of the future system. The participants were invited to come to the laboratory over several days, where a representation of their future problem solving environment had been built (also on the basis of proposed RTCA, 1995 rules). This consisted of airspace maps (where flexible route airspace had been drawn in to represent possible future airspace layouts) and static representations of future radar displays (including aircraft symbols). In advance of the study, all participants were prepared for the procedures of the future (also according to RTCA, 1995), using the kinds of materials that would normally constitute their procedural and policy guidance (for example in this case, faked pages out of "future" air traffic control handbooks and aviation Advisory Circulars).

The four controllers and other practitioners were all gathered in front of one future radar representation, which portrayed the starting situation of the future incident. The initial conditions were given to them, then the anomaly was introduced (e.g. a com failure), from which point they were asked to solve the problem together using the rules of the future. Thus, participants themselves became the engine of action. We decided not to mimic the event-driven and time-limited nature of their operating environment at this stage, so no specific time limit was imposed. The emphasis in this research was exploratory, so it was more important to have practitioners consider as many potential solution paths as possible. Participants could mark up the radar representation to suggest proposed aircraft movements and were encouraged to voice their proposals to the other participants for consideration, which generated protocols based on verbal and motor behaviour that occurred as part of the participants' natural task behaviour. The entire sessions were videotaped and later transcribed, creating a verbal and visual process trace that not only documented the various ways in which the incident could have unfolded, but also captured the cues, tools and rules necessary to address the anomaly in the envisioned architecture.

The dilemma of management by exception uncloaked

Amplifying the Hoogovens experience, this study revealed how the interdependence between when to intervene and how to intervene creates a profound dilemma for exception managers. There turn out to be various pressures not to intervene early in ATC, for reasons that include controller task load, downstream repercussions on system throughput, and even flight safety. But by the time enough evidence is gathered on whether an anomaly really warrants intervention, controllers hardly have an option left to usefully intervene; to contribute meaningfully to a resolution of the unfolding incident. Thus, there are various pressures against intervening late as well, among them flight safety considerations (e.g. the issue of mixed-level control), and controller task load under increasing time pressure. Management by exception traps human controllers in a dilemma: intervening early provides only thin ground for justifying restrictions and creates controller workload problems (and compromises larger air traffic system goals). But intervening late leaves little time for actually resolving the problem, which by then will be well-underway (thereby compromising larger air traffic system goals). In summary, intervening early would be difficult, *and* intervening late would be difficult, although for different reasons. Management by exception seems to put the future controller in a fundamental double bind.

The role of automation

Automating a variety of detection and alerting tasks (such as conflict detection, airspace density calculations and predictions) and providing a controller with computer-generated resolution suggestions does little to alleviate the fundamental dilemma of whether and how to intervene. Asking a machine to do the conflict detection migrates the intervention criterion into a machine, in effect creating a threshold crossing alarm. The typical problem with threshold crossing alarms is that they are set either too early or too late (Woods & Sarter, 1998). When the controller takes over in case of a high threshold, the human intervention may land in the middle of a deteriorating or challenging situation. On the other hand, with low thresholds the alert may come too early to be meaningful. It may become a useless nuisance as the automation flags conflicts without benefiting from the controller's contextual experience, information and knowledge. Such computerisation produces a human-machine system that effectively operates in one of two modes: fully automated (before threshold crossing) or fully manual (after). Controllers might be able to say what is wrong with the machine's decision, but remain powerless to influence it in any way other than through manual take-over (thus easily overloading themselves).

How do we make progress toward a more co-operative human-machine architecture in future ATC architectures? Drawing on experiences in other domains as well as air traffic control (Woods & Roth, 1988; Reason, 1988; Layton, Smith, & McCoy, 1994; Smith *et al.*, 1995; Sarter & Woods, 1992; 1994a; Billings, 1991; 1997), the active partner in a well co-ordinated human-machine team (which in management by exception would often be the machine) would not sound threshold crossing alarms to signal the end of its problem-solving capability. It would instead continuously comment on the difficulty or increasing effort needed to keep relevant parameters on target. The (human) supervisor could ask about the nature of the difficulty, investigate the problem, and perhaps finally intervene to achieve overall safety goals (Woods, 1992; Woods & Sarter, 1998).

These types of co-operative interaction also specify the kinds of feedback that decision support in ATC (such as conflict detection and resolution advisory technologies) would require. For example, the machine partner would have to show when (and why) it is having increasing trouble handling a situation. Displays must be future-oriented to highlight significant sequences that reveal what could happen next and where (Woods, 1994b). This is consistent with other findings of supervisory control demands in dynamic domains: the more a supervisor is distanced from the details of his monitored process, the more his judgements, assessments and decisions will have to be about the future (see Brehmer, 1991). Displays should also be pattern-based, enabling controllers to scan at a glance and quickly pick up expected or abnormal conditions relative to airspace loadings or conflict areas (see Klein, 1993).

In order to build such a co-operative architecture (and inspired by earlier supervisory control work), we should start with determining what levels and modes of interaction will be meaningful to controllers in which situations. In some cases, controllers may want to take very detailed control of some portion of a problem, specifying exactly what decisions are made and in what sequence, while in others the controllers may want to make very general, high level corrections to the course of events. We have begun to explore some of these levels with our studies. In one situation, controllers suggested that telling aircraft in general where *not* to go was an easier (and sufficient) intervention than telling each individual aircraft where to go. These ideas also begin to specify future system requirements (for instance, the ability to communicate to aircraft the unavailability of a piece of airspace because of a particular problem in it).

Conclusion

What seemed like an easy assignment to future air traffic managers ("wait for enough evidence and then decide how to intervene") in fact turned out to be an extremely hard cognitive problem. Their decision to intervene is an intricate trade-off between multiple interacting goals that are simultaneously active (e.g., separation safety, system throughput, controller workload, user economic concerns, etc.). In practical terms for the future system, early interventions are likely to create various misunderstandings with local airspace users who are uncertain about the reasons for sudden restrictions or revisions. Early intervention can also create problems downstream (in terms of throughput, efficiency, backlogs) and controller workload. Late intervention would leave more time for gathering evidence: the exception manager can establish with more certainty that problems are indeed afoot and which aircraft are going to be affected. But every second spent assessing a situation will have been lost resolving it. Management by exception traps human controllers in a dilemma: intervening early provides little justification for restrictions and creates controller workload problems (and compromises larger air traffic system goals). But intervening late leaves little time for actually resolving the problem, which by then will be well-underway (thereby compromising larger air traffic system goals). In conclusion, intervening early is difficult, *and* intervening late is difficult, putting the exception manager in a double bind.

It may also be problematic for air traffic management by exception to become smarter with experience. Consultations with practitioners inside and outside the studies reported here indicate that late interventions may frequently trigger a "that was close" reaction, and drive a swing back over to early interventions. This will disrupt learning, since early interventions resolve anticipated problems even if there was no actual problem developing. They eradicate any evidence on whether the intervention itself was actually warranted, possibly allowing slippage back to late interventions until "it was close" once again, and recycling the pattern.

When computerised decision support provides threshold crossing alarms (e.g. exceeding pre-set closure rates or dynamic density figures) to flag exceptions, they do nothing to resolve the fundamental double bind between difficult early and difficult late interventions. To assist the human supervisor in making the call to intervene, the machine portion of a more co-operative human-machine architecture would instead inform the human about the difficulty or increasing effort needed to keep relevant parameters on target, allowing the supervisor to probe the nature of the difficulty and assess the requirement of his involvement.

Cracking the envisioned world problem

Despite the intentions of those involved in ATC improvements, many developments remain fundamentally technology-centred. Developing the technology (automatic dependent surveillance, conflict probes, digital communications) remains the primary activity around which all else is organised. The focus is on pushing the technological frontier; on creating the technological system in order to influence human cognition or human activity. These efforts are likely to produce ideas (for instance that controllers are effective as exception managers) which are based on generous assumptions about human performance. Similarly, these efforts are likely to produce computerised support that is not co-operative from the human controller's perspective.

There are ways to crack the envisioned world problem and to shift the attention away from solely developing enabling technologies. For example, through future incidents we can become able to create test situations that map onto the target (the future) situation in cognitively accurate ways. These are test situations where future system rules and future human roles are exercised in the face of timeless challenges to system integrity (the non-com or depressurisation scenario in ATC, for example).

Human factors' leverage for change decreases as development time goes on, and human factors researchers typically face trade-offs in playing a role in system development. But this is not necessarily a trade-off between (1) waiting for a fully developed system in order to get valid results versus (2) going in early and getting invalid results as a consequence of empirical intractability. The trade-off is instead between generating valid results that can provide useful input to system development early on and generating valid results that can't because they come too late. Validity derives from the extent to which problems-to-be-solved in the test situation represent the vulnerabilities and challenges that exist in the target world, and the extent to which real problem-solving expertise is brought to bear by study participants. Relying on these sources, we can begin to extract empirical human performance data from the future.

9 Modern Flight Training - Managing Automation or Learning to Fly?

JOHAN RIGNÉR AND SIDNEY DEKKER*

Royal Institute of Technology, Stockholm, Sweden
*Centre for Human Factors in Aviation, Linköping Institute of Technology, Sweden

Introduction

Although automation has fundamentally changed the roles of people on the flight deck, it has not reduced the need to invest in human expertise (Sarter, Woods & Billings, 1997). Rather, it has changed this need profoundly. However, few know where and how precisely to invest in human expertise so that it can begin to match the novel demands of the automated workplace. What we do know is that flying a glass cockpit airliner is essentially a management task. Today's pilot has to be able to manage and co-ordinate activities across multiple cognitive agents or crewmembers (both humans and machines). This includes the ability to plan ahead and anticipate future events, the ability to distribute workload over time, maintaining mode awareness, monitoring, and carrying out other tasks characteristic of a supervisory role. The question is, where do we start teaching these kinds of skills, and how? The airline industry is struggling to find answers. Meanwhile, much training at the basic level remains virtually unchanged. Airlines are under renewed pressure to find qualified pilots to fill recent vacancies, and have few options but to rely on methods that worked before (Nash, 1998), even though obsolete relative to the demands of the automated flight deck (GAO, 1997; NRC, 1997). New Scientist (1996) captured the problem by asking: "Are crew training and cockpit technology pulling in opposite directions?"

Difficult tasks, diffuse investments

The investments in training for automation have not been equal across the training board. Transition training (from non-glass to glass cockpits) has received ample research and airline attention. Beginning with Earl Wiener's B-757 work in the late eighties (Wiener, 1989), it recently culminated in a cross-European research effort on transition training and flight safety (ECOTTRIS, 1998). Transition training is in no way a solved problem, however (see for example Chidester in this volume). But it is safe to say that airlines themselves remain the chief beneficiary of many of the research and development investments, such as training to prevent automation surprises (Amalberti, 1997). This leaves most of the phases of flight training that lay the groundwork for a pilot's competence behind. For a variety of reasons, these phases have a hard time integrating automation and modern flight management meaningfully into their curricula.

One of these phases forms the topic of this chapter. We discuss *ab initio* training, the phase in which new recruits are brought up to commercial pilot status in about 200 hours of flying. Due to a variety of constraints (many of them regulatory), training in this phase has basically not changed since the Second World War. But this phase would offer a fertile opportunity to prepare young pilots for their future flight management task on top of basic flying skills.

We don't reflect on this phase in a regulatory or organisational vacuum. The civil aviation training industry is trying to readjust to the new demands, while finding itself in a force field governed by legislative and economic constraints and customer demands. Moves to harmonise the training of pilots across many European countries are currently being implemented. The so-called JAR-FCL's (Joint Aviation Regulations, Flight Crew Licensing) will affect all stages of pilot training and certification, from *ab initio* to multi-crew jet type ratings. This has enormous ramifications for how operators and flying schools must organise and manage their pilot training. We try to follow proposed JAR-FCL as well as we can, bearing in mind that JAR-FCL is not only huge, but that it is also a living document. This means that proposals discussed here may have changed or amended in some way by the time this is read.

Introducing pilots to automation

The practical handling of automation is generally introduced late in a pilot's curriculum (see e.g. Sherman, 1997), something JAR-FCL is trying to readjust, if modestly. Currently the first brush with automation often coincides with a pilot's introduction to multi-crew cockpits and jet transport aircraft, i.e. when joining an airline. This pressurises an airline's training resources and the individual pilot's capabilities. Failure rates during initial airline training, type rating and conversion training can be high (and are expensive). Other documented consequences include the development of misconceptions and simplifications in the knowledge and use of automation (Wiener, 1989; Sarter & Woods, 1992; 1994; Sarter, 1995). Training programmes often emphasise those parts that are easiest to learn. In an effort to deal with the complexity and volume of topics and skills to be learned, there is a tendency to focus on automation usage "recipes". The rest is for the pilot to learn on the line. One result is that it often takes more than a year of line flying on the one aircraft type for a pilot to find his way in the tangled web of automated modes and levels.

Knowledge issues

If the goals of flight education are to make pilots able to transfer their knowledge (from the training situation to the airline environment), so they can manage both routine and novel situations, training methods that rely on reproductive memory do not make the grade.

Against a backdrop of much routine work and applications of rote procedures (which are either in the head or in the world), cognitive activity on the professional flight decks ebbs and flows. Periods of lower activity and more self-paced tasks are interspersed with busy, externally paced operations where task performance is more critical (Rochlin *et al.*, 1987). These higher tempo situations create greater need for cognitive work and often require novel applications of knowledge. At the same time they often create greater constraints on cognitive activity (e.g. time pressure, uncertainty, exceptional circumstances, failures and their associated hazards) (Woods *et al.*, 1994). When it comes to the use of knowledge in these kinds of situations, the linkage or mapping between conceptual knowledge and its application is not straightforward, but rather complex and irregular (Feltovich *et al.*, 1993). Accidents and incidents where automation complexity, mode confusion, basic flying skills and crew co-ordination all interacted and jointly contributed to the sequence of events, show that in reality many concepts, in interaction, are pertinent to one instance of knowledge application (NTSB, 1986; 1994). Isolated, passive knowledge components may not activate and combine in the required ways, especially when the tempo and volume of human problem-solving activities become driven by events and activities in the environment.

We often assume that knowledge shown in one context (for example: a written exam) is retrievable in any other context. But in reality there is a dissociation effect: knowledge learned and shown in one context is not necessarily applicable to any other context (Woods *et al.*, 1994). Knowledge can also remain inaccessible or inert if the ways in which knowledge needs to be combined and applied do not resemble the way in which it was acquired. In the words of Feltovich *et al.* (1993), knowledge will be more readily available for later use if the settings, cognitive processes, and goals active at the time knowledge needs to be used resemble those that were active when knowledge was acquired. Knowledge for managing complex, dynamic domains, then, must be looked upon as a tool, something that is constructed in interaction between a mind and situations calling for action.

The implications of these knowledge issues for pilot training are as follows:

♦ Separating automation from other technical topics in a theoretical knowledge curriculum misses the point. The flight deck of a modern airliner *is* automated. No matter what action pilots take during a normal flight, (s)he is likely to interact with an automated system, be it the flight management system, the flight controls, etc. In other words, treating aircraft automation (and its sub-components such as autopilot, flight director, etc.) in isolation and on par with other aircraft sub-systems (e.g. hydraulics) is inconsistent with operational practice and will be ineffective in transferring automation knowledge to the actual flying task.

♦ The distinction between technical and non-technical skills has eroded. Operating the aircraft (via its automation) is as technical (knowledge of modes, levels, transitions) as it is non-technical (e.g. keeping the other pilot in the loop relative to instructions given to the computer, to aircraft behaviour and other expected feedback).

♦ Training "bandaids" to fix automation difficulties only once they arise in transition training are rarely completely successful. Telling pilots not to fall into certain automation traps does not mean they will not fall into them. Exhortations to be careful, although present in most training, generally do not work. They come too late in the learning process and must be applied successfully in contexts (dynamic, complex, event-driven, high stakes) that are fundamentally different from the teaching context. They also imply that pilots fall into traps because of a lack of motivation and that exhorting them to "try harder" is helpful and meaningful.

♦ *Ab initio* programs that prepare pilots for generic supervisory skills (delegation, monitoring), workload management and team play on top of basic flying skills will be more effective in transferring knowledge from the basic training phase to the actual airline work environment.

Ab initio training and automation

The training and certification of pilots at the basic level has not really changed since the fifties (Lehman, 1998). Many believe that training methods and flight crew licensing standards for these kinds of pilots have been outrun by the rapid technological developments of the past two decades. In the wake of the Birgenair crash which occurred to an automated airliner off Puerto Rico, Flight International (November 1996) opined: "The Birgenair report holds back from blaming the airline. It blames the aviation authorities which allow crews to be licensed using training systems and standards which are hopelessly outdated in the world of high-performance, highly automated aircraft".

The finding that training systems and standards might be outdated - at least in critical parts - is echoed by many in the industry. According to Graham Hunt (1991), crew education and training in the Western world have hardly changed since the mid-1950s, except for the development of full flight simulation at the airline level. At the *ab initio* level changes have occurred only in the margins, not in any fundamental sense. Radical departures from current *ab initio* training practice are suggested by some (e.g. Lehman, 1998), who suggest that training future pilots on single engine aircraft with low power and low wingloading does little to prepare them for their future role as airline pilots.

Automation in the ab initio curriculum

Proposed JAR-FCL legislation signals that an integrated airline pilot course should train pilots to the level of proficiency necessary to enable them "to operate as co-pilot on multi-pilot, multi-engine aeroplanes in commercial air transportation" (JAR-FCL 1.160 and 1.165). This means that flying schools could focus solely on turning *ab initio* pilots into multi-crew, automated cockpit team players. But the regulations don't go very far in suggesting how. The main regulatory change is the introduction of new topics (e.g. automation) rather than suggesting how the teaching and training should be performed and adapted to the new management work on the modern flight deck.

The issue is also that automation is treated as a separate sub-component. The autopilot and flight management system take their respective places among the hydraulic system, electrical system, etc. Although this may be a truthful reflection of an aircraft from an engineer's standpoint, such parsing of knowledge misrepresents operational reality. In actual airline flying, automation pervades every flying task, including flight planning, lateral and vertical navigation, performance and fuel management, and is nowadays itself the window through which pilots interact with most if not all other subsystems (Billings, 1997). Empirically, pilots treat the automation as a third crewmember, an animate agent capable of using knowledge in the pursuit of its own goals (Sarter & Woods, 1994a). If automation is looked upon as a third crewmember, then teaching future pilots compartmentalised knowledge about the flight management system or autopilot as sub-systems will add little to their ability to manage the automation in normal and abnormal situations.

The limited flying time received during *ab initio* training is mostly gained on aircraft whose level of automation has little resemblance to that of the pilot's ultimate working environment. One justification for this training is that basic flying skills will continue to be essential. The context in which they are trained could however be changed to better reflect their changed role in airline flying, where they are left only for certain portions of the flight as well as for recovery from unusual attitudes which ironically often have their roots in automation-confusion (FAA, 1996).

This also means that technological "catching up" of training aircraft is only one way forward. The mapping of the training situation to the target situation (i.e. the future work environment) does not necessarily have to rely on face validity (making small aeroplanes look like big ones). The critical mapping is cognitive, where fundamental skills and knowledge learned in *ab initio* training become useful and applicable to flying the line.

Establishing the environment of the future

With JAR-FCL's allowing flying schools to train pilots specifically for multi-crew large aircraft, the goal during *ab initio* training must be to establish a training environment (or context) as similar to the environment on the modern multi-crew jet transport aircraft as possible. This environment could be established from the first day of training where automation is not treated as a separate subsystem as any other but rather as a system with major repercussions for the overall management and conduct of any flight. Even without advanced automation in training aircraft the future performance of the student could benefit from (repeated) learning that manual flying is only one among other modes of flying (and building a mental model with that in mind) while practising the basic flying skills.

In practice this could imply the introduction of concepts such as design and automation philosophy, automation policies etc. during initial training. Various approaches towards automation taken by airlines could be discussed, the consequences and differences between moving and non-moving throttles explained etc. One obstacle to overcome is the fact that many of the instructors at Flying Training Organisations (FTOs) have never been professionally exposed to modern flight decks or themselves received any instructions on the differences between various design philosophies.

Training for the test

Another important factor that influences the transfer of knowledge and value of basic training is how tests are performed. Test requirements and methods are embedded in flight crew licensing legislation. In reality, training organisations and pilots themselves will train for the test, and not necessarily for the (future) task. This is especially the case when a large number of subjects are to be learned during a short period of time (which is often the case during *ab initio* training). To be relevant (and to ensure relevant training), tests on automation must require the same kinds of successful cognitive performance as is necessary to conduct safe line operations with modern aircraft. These concerns are increasingly recognised and expressed by the airline industry (see e.g. Bent, 1997).

Conclusion

Treating automation as a separate subject may misrepresent its place in the curriculum of modern flight training. It is rather a non-subject in the way that automation *is* the environment and the context in which the other (non-technical) subjects should be taught.

At the *ab initio* level, the impact of automation on the flying and management task in the modern cockpit should be addressed. Despite increasing regulation in this area, it is still very much up to individual flying training organisations to realise this and design training programs accordingly. This chapter has proposed several ways in which automation and preparation for future supervisory control work in cockpits could be meaningfully integrated in those stages where the groundwork for pilot competence is laid.

10 Introducing FMS Aircraft into Airline Operations

TOM CHIDESTER

American Airlines, USA

Introduction

Billings (1997) described the introduction of the 757/767 as beginning a fundamental shift in aircraft automation - it was the first of a generation of Flight Management Computer (FMC) aircraft. These aircraft integrate area-navigation functions with aircraft performance management, allowing computer entries to guide or control the aircraft in four dimensions, and integrate warning and alerting systems for mechanical problems. Those capabilities were built upon in subsequent aircraft, including the A300-600, A310/320, B747-400, B737-400 through 800, F100, MD-88, and MD-11. Each new aircraft of this generation introduced greater navigation capability or system integration through computer-based management systems.

This fundamental shift began nearly 20 years ago, yet airlines continue to make significant changes in policy, procedure, and training for these aircraft. Those changes are in response to human performance issues growing from the automation of aircraft functions. These two decades could fairly be described as a struggle to integrate a new generation of technology into our operating environment. In fact, the U.S. Federal Aviation Administration (FAA) Human Factors Team (FAA, 1996), a group of representatives from FAA Certification and Standards, National Aeronautics and Space Administration (NASA), and European Joint Aviation Authorities (JAA), and selected technical advisors, described a number of unresolved issues in their report. The team identified vulnerabilities in pilot management of automation and situation awareness. These included understanding the capabilities, limitations, modes, and operating principles of automated flightdeck systems and choosing levels of automation appropriate to flight situations.

This chapter reviews the range of issues identified to date as human performance consequences of aircraft automation in this generation of aircraft within airline operations. It also reviews the policy and training changes that have resulted. I will argue that identifying these issues and developing responses are not unique to this generation of aircraft. Issues in the interaction of new technology with the existing operating environment are not and perhaps cannot be anticipated entirely at the design level, even where human factors specialists are a part of the design process. And so, introducing that technology requires a significant assessment and monitoring of its use at the operator level. Rather than offering criticism of design of this or any other generation of aircraft, then, this chapter argues for development and application of human factors expertise at both the design and operation level.

Characteristics of FMS-generation aircraft

Though described as a fundamental shift, the introduction of aircraft with Flight Management Computers (FMC) did not represent the introduction of automation into flight operations. To the contrary, automation of aircraft functions has proceeded in increasing levels of sophistication over time. Consider the graph below of the automation on one carrier's aircraft fleet:

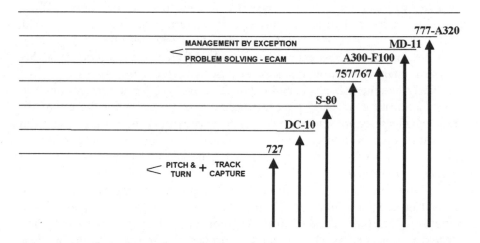

Figure 1: Levels of automation

From this perspective, each generation of aircraft has some level of automation, ranging from a simple autopilot combined with a form of area navigation (Omega or Inertial Navigation systems; ONS, INS) on the B-727, to "system management by exception" on the MD-11. Many of the operational differences among the aircraft operated by this carrier can be described as levels of automation. Its 727s are equipped with a "primitive" area navigation system - the capability to navigate point-to-point without direct reference to radial/bearing/distance from a ground-based facility - and these are now being replaced with GPS-based navigation systems nearing the sophistication of FMS aircraft. The potential for many automation issues can already be found at that level. From the 727, flight guidance and autothrottles were introduced with the DC-10, and a primitive vertical navigation system, PERF, was introduced with the S-80. True point-to-point lateral and vertical navigation (LNAV and VNAV) were introduced with the 757 and 767, and this was the first level to be perceived as revolutionary change by most pilots. Perhaps this is because of the form of data entry and the introduction of complete CRT displays for flight and engine instruments. But again, this was only a plateau - most of the visible automation to this point had been flight path or navigation functions. With the introduction of the A-300-600/A-310 and the F-100, the focus turned to automation of how the pilot controls systems, beginning with automatically-referenced electronic checklists and continuing to a management-by-exception approach to systems on the MD-11, where abnormal on some systems are corrected or dealt with and the pilots notified. The A320 and B777 introduced sophisticated envelope protection systems (though by very different philosophies, with the Airbus restricting inputs that would exceed performance limits and Boeing providing physical feedback of approaching limits).

The point is that all air carrier aircraft have some level of automation - talking exclusively in terms of automated versus non-automated aircraft is misleading. It is more important to talk about the consequences of whatever level of automation is available on an aeroplane. Each has not only a maximum, but also a range of levels from which the pilots may select. Even the MD-11 can be "hand flown" or operated in intermediate, vertical speed and heading modes.

Levels of automation on FMS-generation aircraft

To simplify the discussion, consider an FMS aircraft to have three levels of automation of flightpath control, with the third level its unique advance over previous generations. At the lowest level is *hand-flying* the aircraft without flight director guidance. At this level, the pilot may essentially control all pitch, roll, yaw, and power settings. The pilot establishes control of each dimension by scanning instruments and monitoring the results of inputs to elevator, aileron, rudder, and thrust, respectively. This task is fundamentally the same in an air transport or a single-engine training aircraft.

At a second level, which I will call *mode control*, the pilot controls each dimension by controlling targets on a mode control panel. For example, a desired speed and/or rate of climb is entered and a mode, such as vertical speed, is selected. If the pilot is hand-flying, a flight director will provide pitch guidance on the primary flight display (artificial horizon), and thrust guidance on engine gauges (exhaust pressure ratio (EPR) or engine speed (N1, N2, or N3)). If autopilot and autothrottles are engaged, their servos will manoeuvre the aircraft and throttles to the pitch and power settings that will yield the commanded speed and vertical speed.

At a third level, which I will call *flight management*, the pilot controls each dimension by programming a plan into the flight management computer, including a route of flight, speeds and altitudes at any waypoint, and on some aircraft, arrival time at a waypoint. If the pilot is hand-flying, a flight director will provide pitch and roll guidance on the primary flight display and thrust guidance on engine gauges. If autopilot and autothrottles are engaged, their servos will manoeuvre the aircraft to pitch and power settings that will efficiently accomplished the desired path.

Two analogies may aid the reader's understanding of levels of automation. The first involves the use of sequencers in music. A musician striking a key on a piano or synthesiser is playing a note. A rhythm machine may be used to put together a series of notes or drumbeats into short patterns, which are repeated throughout a piece of music. And a sequencer may be used to combine a variety of notes and patterns into a composition. The key to this analogy lies in what unit is being manipulated - *hand-flying* a synthesiser involves playing notes, *mode control* involves manually selecting patterns, flight management involves programming into a song.

The second analogy is Endsley's (1988) description of situation awareness at three levels - "the perception of the elements in the environment within a volume of time and space, the comprehension of their meaning, and the projection of their status in the near future." Robertson & Endsley (1994) further explicated the levels as (1) awareness of information or *data*, (2) awareness of its meaning relative to operational goals such as position relative to *targets*, and (3) projections of likely future events and scenarios - *strategy* or future path of the aircraft. The concept of levels of automation maps neatly to the manipulation of each of these parameters. *Hand-flying*, like pitch wheels and turn knobs from previous generations of autopilots, directly manipulate flight controls, and therefore, *data* - pitch, power, roll, and yaw. *Mode control* - heading hold or select, altitude hold or flight level change, and speed modes - manipulates *targets* - near term tactical goals. *Flight management* - lateral and vertical navigation - manipulates *strategy*.

The sophisticated reader will recognise that this is a simplification along several dimensions. All the aircraft of this generation have significant automation of warnings and of systems as well as flight path control. For example, all of the aircraft have centralised alerting of mechanical malfunctions or parameter exceedances, and some also will locate and display an appropriate checklist for resolving the problem. All also have some automation of systems - pressurisation, for example, so that pilots may enter the departure and arrival field elevations and the aircraft will maintain throughout flight a pressure differential that is both safe for the structure of the aircraft and habitable for the passengers. There have been incidents involving system automation, such as pilots attempting to complete a checklist for an alerted symptom (generator failure) rather than its unalerted underlying cause (engine operating at a sub-idle speed), just as there have been incidents with automated flight path control.

I have chosen to focus on the automation of flight path control because this is where the majority of the problems or issues in human performance associated with this generation of aircraft have been observed. I have also set aside the discussion of area navigation for now, because navigation to latitudes and longitudes from a known position is not unique to this generation of aircraft. A distinction may also be drawn between engaging the autopilot and hand-flying to guidance whether from mode control or flight management. Some airlines have found this distinction useful and some have not. I will discuss this in more detail later in the chapter. Why, then, simplify the discussion to three generic levels of automation? Because these concepts tend to capture or communicate effectively the issues identified on FMS-generation aircraft to the pilots who operate these aircraft.

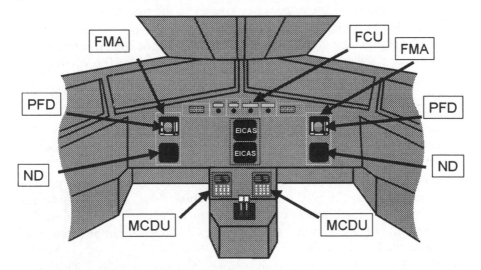

Figure 2: Automated cockpit

Displays and controls on FMS-generation aircraft

One additional point is in order for those unfamiliar with automated cockpits. Figure 2 presents some of the basic displays and controls introduced in FMS-generation aircraft. Each aircraft may call the same unit of this prototypical automated-cockpit by a different name, but in general all share these displays and controls.

The displays and controls introduced by FMC aircraft include:

♦ MCDU: Multifunction Computer Display Unit. This is the data entry, editing, and display interface of the Flight Management Computer. Flight plan and weight and balance information is entered or radio-uplinked through this interface. Information calculated by the Flight Management Computer is displayed to the pilot.

♦ FCU: Flight Control Unit or Mode Control Panel. This is where the pilot implements most level-of-automation decisions, including which dimensions to control and through which devices to execute control. The pilot selects pitch, power, roll, and speed modes, may enter altitude, heading, or speed targets, and decides whether guidance will be provided by mode control or flight management. Flight Director and Autopilot engagement switches also appear on this panel, though each aircraft provides autopilot disengagement switches on the yoke as well.

♦ FMA: Flight Mode Annunciator. This device informs the pilots of the modes engaged by flight guidance or flight management. As a result, this becomes a key display. It tells the pilot whether the aircraft is carrying out his or her control decisions.

♦ PFD: Primary Flight Display. This varies from aircraft to aircraft but always includes an artificial horizon, display of flight director guidance, and glide slope and localiser deviation. On some aircraft the FMA also appears. On some, airspeed tapes, radar altimetry, and even angle of attack may be displayed.

♦ ND: Navigation Display - This is a switchable device which may present a moving map display, an electronic horizontal situation indicator (EHSI), or a combination of map and course deviation information. On some aircraft, TCAS, weather radar, or EGPWS terrain information may be overlaid on the map.

♦ EICAS: Engine Indicating and Crew Alerting System - This is a centralised display of system alerts or warnings and engine indications. On some aircraft these displays will present a checklist for resolving an alert or warning as well.

Defining the issues

It is a great reach from this simple description of what FMS-generation aircraft introduced, to the FAA (1996) conclusions of human factors consequences. This section attempts to define the issues growing from the automation of flightpath control functions. The sources are both formal research and feedback from line operations. The issues are therefore both conceptual and empirical. My goal in this section is to highlight those issues identified in human factors research that have been understood and acted upon by operators. This is necessarily less extensive and less theoretical than many of the other chapters in this book - I am attempting to document the influences on current policy, procedure, and training for this generation of aircraft. My intent is to communicate the concepts and actions that are currently being emphasised to the pilots of these aircraft at major airlines. I will, therefore, use the same of the same language and examples I have used to communicate these issues to line pilots.

Automation as "distance" from primary flight controls

Automation is a matter of levels, but it also can become distance between the pilot and the flight controls. Billings (1997) described this as a movement of pilot action from inner to outer control loops, cf. Figure 3.

The first aircraft flown by most pilots has only manual controls - hands and feet operate flight controls, and the controls provide physical feedback of what the aircraft is doing. This is fun, but the pilot is always busy controlling the aircraft. Automation comes between the pilot and that active, physical workload - and its feedback. As levels of automation are added, the feedback changes from active to passive. The aircraft communicates feedback through monitors and instruments. This can be beneficial or detrimental, because "distance" from the flight controls decreases physical workload. Using automation can free pilot attention to care for other necessary activities. But, this distance can affect situation awareness. Pilots can fall out of the loop, losing awareness of what the aircraft is doing.

Endsley's concept of situation awareness at three levels becomes very relevant here. Each higher-level of automation follows a user-transparent or user-limitable control law to manipulate the data or targets required to meet what has been commanded. So, by selecting vertical navigation, the pilot cedes direct control of vertical speed (though it could be increased by deploying speed brakes, or decreased by constraining airspeed), for example. Similarly, situation awareness relative to each lower level may be diminished while using each higher level of automation. By manipulating strategy through lateral navigation, a pilot may become less aware of targets. By manipulating targets, he or she may lose awareness of data. This point will be important to the discussion later in the chapter.

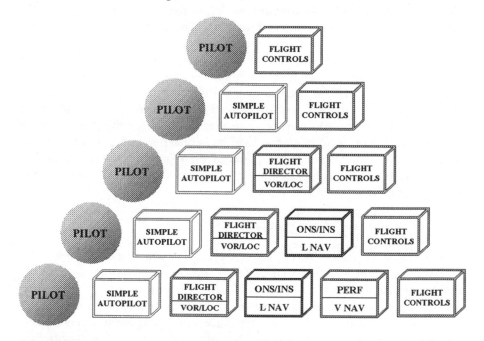

Figure 3: From inner to outer control loops

Workload

Automation by definition is intended to reduce workload. But other benefits come with automation, as well. As aircraft have become more automated they have also become more reliable and efficient. And very clearly, automation has improved navigation and all-weather operation. The measurement of workload was key in the development of this generation of aircraft. In part this was due to the desire to move from a three-pilot to a two-pilot operation - automation of systems and control would be necessary with a reduction of the number of pilots in the cockpit (McLucus, Drinkwater & Leaf, 1981). As these aircraft were introduced, one of the first controversies surrounded whether workload-reduction was adequate in the line environment. Wiener (1989), after conducting a program of survey research, suggested that this generation of aircraft tended to provide clumsy automation. Workload tended to be reduced in situations where it was already low and increased in situations where it was high.

Wiener, Chidester, Kanki, Palmer, Curry & Gregorich (1989) illustrated this point in a full-mission simulation study. The simulation was designed to assess whether the actions of pilots in an FMS-generation aircraft would be any different from pilots in a traditional aircraft, if both were confronted with the same problem. Would their workload be the same? Would they communicate in the same way? Would both deal with the problem successfully? Twenty-two crews were given an identical simulator scenario and aircraft system problem, half in a DC-9-30 and half in the MD-88. The scenarios were flown in a full-mission format (Ruffel-Smith, 1979), with realistic ATC and company frequency communications.

The scenario simulated a flight from Atlanta (ATL) to Columbia, South Carolina (CAE) during instrument meteorological conditions. As the crew descended into CAE, they received a warning of low oil pressure on one generator drive unit (CSD). A check of indications showed the generator operating normally and the temperature to be only in the caution band. The checklist allowed the crew to continue to operate the generator, while monitoring for any further temperature increase. At this point, almost all of the crews elected to continue the approach, but initiated a start of the auxiliary power unit (APU). As part of the scenario, the APU would not start on the first attempt, a fact usually noticed during the final approach segment. With the aircraft on short final, a light-twin aircraft was reported to have missed the approach. At decision height, none of the crews were able to see the runway and all initiated a missed approach.

Almost immediately, they were instructed to turn left to a heading of 060 to intercept the CAE 030 radial, and climb to 5,000 feet, with holding instructions to follow. This new, non-published missed approach was explained by the smaller aircraft enroute to the published fix. When each crew asked for holding, they were told to hold south on the 030 radial, 20 DME fix with 10 mile legs. This manipulation was important to the study. Using the published missed approach, the MD-88 crews had the advantage of their missed approach course being fully programmed - they needed only to go-around and engage LNAV. With the change, they have no more course guidance on the missed approach than the DC9-30 crews, unless they reprogram the FMC. During the missed approach, the CSD temperature began to climb, and the CSD failed if not disconnected prior to entering holding.

Flying this scenario in the DC-9-30 was straightforward – it could be accomplished by hand-flying or mode control. The pilot flying (PF) must turn to the assigned heading, climb, and level off. The pilot not flying (PNF) had to tune and identify the CAE VOR, because both navigation radios had been on the ILS. Then, one pilot set the 030 radial into the course window and the PF had only to turn right into the holding pattern on arrival at the holding fix. Only a few questions of technique needed to be resolved - did they use the autopilot? the VOR/LOC modes? But each crew also had to monitor and deal with the CSD. As straightforward as this situation might seem, the workload was high, and some of the DC-9 crews made mistakes.

Flying this same scenario in the MD-88 provided the crew with additional options for flying the missed approach. A level-of-automation decision had to be made. As on the DC-9, it could be hand flown, or flown in heading select and vertical speed to the holding fix and altitude. Alternatively, the PF could turn to the intercept heading while the PNF immediately programmed the CAE 030 radial, the 20 DME fix, and the holding pattern, allowing the use of LNAV. Or, some combination of these methods could be used, such as flying to the holding fix while the PNF dealt with the overheating CSD, then programming the holding pattern to alleviate the workload of flying the aeroplane. Assuming that each step was performed correctly, any of these methods would work acceptably well.

But a small number of crews on the MD-88 made some programming mistakes which escalated into larger problems on the missed approach. A typical example looked something like this. As soon as the missed approach was initiated, one pilot selected LNAV, which started the aircraft on the ground track to the published missed approach. Moments later, ATC issued the new heading and altitude, causing one pilot to select Heading Select and Flight Level change – a necessary move to a lower level of automation. Then, the PNF requested the holding instructions and began to program the FMS to show the holding pattern. During this programming, the PNF would make a mistake, resulting in an incorrect depiction of the holding pattern (this mistake is very common and involves inserting the 030 radial of the holding fix as the *holding radial* rather than the *inbound course*). This would often require discussion with the PF to correct the error, and about this same time, the aircraft would arrive at the holding fix and the undetected overheating CSD would fail, resulting in display resets and warnings. At this point, the crew must fly the aeroplane, handle the system failure, and decide where to divert - and the programming necessary to use their chosen level of automation has not been completed.

So, a small number of the MD-88 crews made some serious mistakes, and comparable mistakes happened during the missed approach on the DC-9. What were the overall results of the study? First, there were no significant differences between the two fleets in the performance of the crews. This is a very important point - crews on both fleets completed their scenarios safely.

Mistakes were made on both aircraft. But there were several important differences:

♦ When given a standardised measure of workload (Hart & Staveland, 1988), MD-88 crews who completed the scenario reported significantly higher workload. The crews of the *more* automated aircraft described the scenario as being more mentally and physically demanding than did the crews of the less automated aircraft.

♦ MD-88 crews took longer to complete the scenario. A reasonable, though unproven, hypothesis is that they took more time to deal with the higher workload they were experiencing.

♦ MD-88 crews communicated twice as much during the abnormal period. The number of verbalisations - questions, statements, commands - by MD-88 crews was twice the number among DC-9 crews.

♦ The dominant pattern of communication in the DC-9, a Captain's command or instruction followed by a reply from the First Officer, changed significantly in the MD-88. The most frequent communication became a Captain's *question* followed by a First Officer response. Questions indicate uncertainty. What was less certain in this scenario on the more automated aeroplane?

This study was the first attempt to document systematic differences in how pilots respond to demanding situations that might be related to the automation available on their aircraft. It raised more questions than it answered, but it did point out some potential traps that can occur in this generation of aircraft. When an abnormal event such as an aircraft problem occurs, pilots must decide how to use the automated systems that are available. Control of workload is either a large factor in this decision, or an unexpected consequence of the choice.

This study, when combined with operator experience, suggests a conceptual framework for workload control for this generation of aircraft. As with the "distance" issue discussed above, there is a complex relationship between automation and workload on FMS-generation aircraft. Consider Figure 4.

In a normal situation, pilot workload decreases to the highest level of automation available on the aircraft. This is the intent behind the design of the systems. However, in an abnormal situation, high workload can come from hand-flying the aircraft, *or* trying to program the FMS in the context of other demands. The higher workloads are at the extremes of automation. Intermediate levels of automation probably relieve workload the most in an abnormal or emergency situation. In an abnormal situation, pilots may need to move up from hand flying *or* down from a level requiring programming, in order to reduce workload. In an abnormal situation, a level-of automation decision becomes critical.

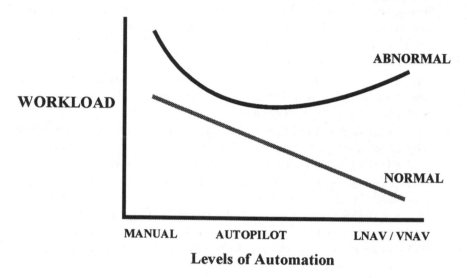

Levels of Automation

Figure 4: Automation and workload

Mode awareness

Based on a series of investigations of pilot interaction with automation on FMS-generation aircraft, Sarter & Woods (1995) identified a range of *automation surprises*. These are situations where crews are surprised by actions taken or not taken by an autoflight system. The opposite of this, which has been called *mode awareness*, is recognition of the current status of all autoflight systems and an understanding of its implications. Consider the following examples discussed by the Air Transport Association Subcommittee on Automation Human Factors (ATA, 1998).

Pilots of MD80 aircraft and its variants have reported altitude deviations following entry of a speed restriction into the Flight Guidance Control Panel. Typically, the aircraft was operating in Perf Cruise mode when ATC restricted speed. The PF dialled in the speed or mach setting, then engaged Speed Select – moving from flight management to mode control. The autopilot then complied with the speed, but began a spontaneous climb or descent. These events result from the design assumptions of the autoflight system software. Exiting Perf Cruise, the autopilot needs both speed and altitude targets. The pilot has provided the speed target. The altitude target is obtained by examining the aircraft's current vertical speed. Because Perf Cruise allows 150 ft. of altitude variance to minimise pitch/power changes, the aircraft is often in a shallow climb or descent. In these events, the altitude target has defaulted to vertical speed. The pilots expected it to default to altitude hold, as the aircraft had previously been at cruise. The obtained mode was annunciated on the FMA, but undetected by the pilots for some period of time.

On F100 and A300 aircraft with the autopilot engaged in certain flight envelopes, pilot elevator input forces are sensed as unwanted inputs and are trimmed out by the autopilot. No force disconnects are implemented, and autopilot trim in the opposite direction of input is not inhibited as on some Boeing aircraft, for example. This has resulted in events that were perceived or described by pilots as runaway trim. For example, according to a report submitted to the Aviation Safety Reporting System (ASRS), the pilot of an F100 aircraft did not disengage the autopilot when initiating a visual approach, but believed he had. Stab trim then moved to both full nose-up and full nose-down positions in response to elevator inputs and configuration changes. The autopilot was attempting to maintain the last target selected on the MCP, which was altitude hold. This sequence ended when torque forces applied to the manual trim wheel by the PNF were sensed as a fault, disengaging the autopilot. Importantly, the pilots believed the autopilot to have been disengaged throughout the manoeuvre. A similar event resulted in an A300 accident at Nagoya, Japan, when the FO/PF engaged TOGA mode, but attempted to stay on the glide slope with forward elevator pressure. Nearly full nose-up trim was obtained as the autopilot sought go-around pitch. The aircraft eventually pitched into stall attitude when the autopilot was disengaged, go-around thrust applied, and flaps retracted. In both of these events, the pilots expected action counter to the underlying software design, though the modes engaged were annunciated on the FMA. The A300 has been modified since the accident so that high control column force will disconnect the autopilot.

Special FMS (RNAV) arrivals and approaches may be flown with lateral and vertical navigation engaged, and the lowest altitude restriction on the profile set in the altitude window. The VNAV mode honours all intermediate restrictions, but will not descend below the value inserted into the altitude select window on the mode control panel. On the B757/767, pilots have reported events in which the aircraft was falling behind (above) VNAV path and the pilot selected FLCH or VS to expedite the descent. Altitude protection is forfeited for the intermediate restrictions by this action. This is a serious error if the pilots are not monitoring the performance of the aircraft at the intermediate restriction. Unsatisfied with the performance of VNAV (flight management), the pilot has moved down a level to mode control. Unfortunately, he or she selected a mode that meets the descent goals but removes altitude protection, and does not recognise that trade-off in real time.

Other examples can be described for virtually every model of FGS, FMS, or GFMS-equipped aircraft. Each of these events can be described as a failure of mode awareness. If the aircraft was in a mode other than what the pilot expected, this may be a simple failure to check the FMA. But some of these events represent underlying misunderstandings of the functioning of particular modes by the pilot - at least in real time - which would appear to be something more. The FAA (1996) concluded that mode awareness represents a significant vulnerability of flightcrew situation awareness. Sarter & Woods (1995) suggested these kinds of surprises flow from increasing autonomy and authority of autoflight systems, combined with inadequate communication of system status. This requires definition of several terms:

♦ Autonomy. The autopilot may initiate actions previously authorised by the pilot or ancillary actions required to carry out a command. Thus, the autopilot can initiate actions without immediately preceding operator input, and change modes as it carries out those instructions. Autonomy is a factor in both the S-80 and B757 examples. In both cases, the pilot authorised more than he or she intended, a pitch change in addition to a speed change and deletion of intermediate altitude restrictions, respectively.

♦ Authority. The autopilot may modulate or override pilot input. The resulting behaviour is determined by the system designer and not the system operator. The F100/A300 example involves the autopilot having pitch authority over pilot inputs. Having engaged the autopilot in a pitch mode, it assumes responsibility for pitch to meet its target and will trim out the pilot's input to the elevator. Automation authority increases with varying degrees of envelope protection.

♦ Envelope protection. Ability of the autoflight system to detect and alert, prevent, or recover from pre-defined unsafe aircraft configurations (e.g. stall). Once an undesired configuration is approached or detected, the automation may have the power to override or limit the pilot input. Thus, protection may be broken into "Hard and Soft" envelopes.

> **Soft Envelope**: Protection that the pilot can override. For example, on the F-100, speed envelopes are protected. When the airspeed falls five knots below reference speed, the throttles set minimum thrust to maintain this speed. Thirty-three lbs. of force are required to overcome throttle pressure.

Hard Envelope: Protection that the pilot cannot override. For example, on the A300-600, Alfa Lock prevents retraction of the leading edge devices at high angles of attack and low airspeed.

This perspective suggests that the greater the autonomy and authority of the autoflight system, the greater the likelihood of a mode awareness failure or automation surprise. How do pilots maintain mode awareness? Presumably, they would start with the same techniques they have always used to maintain situation or position awareness - a basic scan. But, Sarter & Woods concluded that pilots' instrument scan changes on FMS-generation aircraft, moving to one of confirming the result of inputs. In that process, pilots sometimes fail to detect data and target deviations in a timely manner. Consider Figure 5, which illustrates the traditional "T" scan:

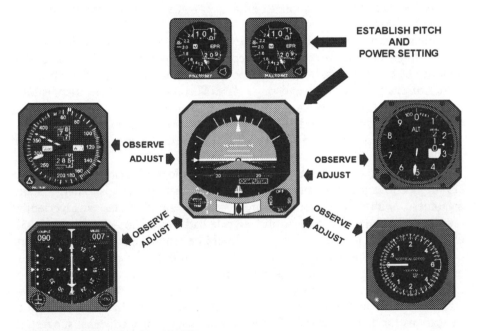

Figure 5: The traditional "T" scan

Basic instrument training for most pilots teaches an instrument crosscheck centred around the PFD or artificial horizon. Pilots establish an approximate pitch and power setting to obtain their desired target, then crosscheck other instruments. The pilot constantly scans altitude, airspeed and heading. *This is a deviation-correcting crosscheck - the pilot is looking for data that does not fit his or her goals, so that corrective adjustments can be applied.* This crosscheck would increase in intensity during critical events such as while intercepting a course and glide slope. It would expand to the flap gauge and gear indicator lights while initiating an approach, for example. Note that this information is distributed across instruments.

On FMS-generation aircraft, instrumentation begins to change and this has some effect on scan. Information is consolidated onto fewer instruments or displays, tending to make the scan more localised. Flight director commands become almost always present and highly accurate - if the pilot places the aircraft symbol on the command bars on the PFD, the selected targets are almost always obtained. Flight Mode Annunciations become increasingly critical. If the flight director or autopilot is not in the right mode, the commands or controls will not seek the desired target. Each of these factors reduces the motivation to scan *data* - the elements making up the standard T scan.

Pilot attention is a limited resource. It is drawn toward what is important. On FMS-generation aircraft, this becomes the selection of modes, the annunciation of modes, flight director commands, and lastly the absolute values of pitch, power, roll, and yaw - data. This *confirmatory* scan, then, is one that attends toward mode selection/annunciation and flight director commands and away from data. *This cross-check seeks to confirm that annunciations are consistent with intentions, which is logically different from identifying and correcting data deviations.*

Does this type of scan fail to detect significant deviations? The FAA (1996) documented a number of events in which pilots lost energy state awareness with the autopilot engaged but autothrottles inadvertently disengaged, failing to recognise the aircraft had moved into a high angle of attack with a low power setting. This should be readily apparent from a scan of data - pitch is increasing, airspeed is falling, altitude is constant, and power is well below target. But in these incidents, the pilots did not recognise the underlying problem for an extended period of time. This would support the contention that attention, scan, and situation awareness move towards the indications of higher levels of automation when engaged, and suggest a vulnerability of this approach. When the pilot makes a mode selection error, misunderstands the implications of a mode, or encounters an unexpected mode transition, scanning the FMA will correct only the first. Continued scan of both the FMA and data is necessary to the latter two.

Paradoxically, Billings (1997) concluded that pilots often detect that the autoflight system is not performing as expected only after it has exceeded the target. In other words, only when it is observable in data. (The resolution of these viewpoints may be found in alerting systems – altitude deviations and approaching stall or terrain produce aural warnings. If this view is correct, the pilots are not detecting deviations by observing data.) Thus, a confirmatory scan does not appear to be sufficient to recognise when the aircraft is not performing as expected. With a de-emphasis on scan of data, corrections will be significantly delayed.

Endsley's model of situation awareness can also be applied to this issue. If in recognising the criticality of mode awareness, pilots bias their scan toward the FMA, what will happen to data awareness? We would predict the kind of delayed recognition described in this section.

To ensure they maintain mode awareness, then, pilots must deliberately scan the FMA to determine whether autopilot and autothrottles are engaged and in what modes. The ATA Subcommittee (1997) described this as analogous to the schedule bidding process. A mode selection is a *bid*, which must be compared with its annunciated *award*. Then aircraft performance - data - must be continuously scanned for *reassignments* - deviations from selected and annunciated targets.

Synthesis of the issues at the operator level

Defining the issues within one airline

Ewell & Chidester (1994) attempted to synthesise the issues identified in the research described above with feedback from check airmen instructing and evaluating on FMS-generation aircraft. Their discussion was most influenced by the workload issues described by Wiener (1989), but anticipated some of the mode awareness issues raised by Sarter & Woods (1995). While recognising that FMS-generation aircraft are both more reliable, capable, and empirically safer than previous generations, they described the following tendencies among pilots on these aircraft:

♦ *Pilots operating more automated aircraft describe a transfer of some aircraft control to PNF.* Beginning with Flight Guidance System-equipped aircraft, which pilot tunes navaids, selects courses, and sets altitudes becomes ambiguous (Wiener, 1989, 1993). And with the introduction of lateral and vertical navigation, the PNF *can* (but should not) make changes in flight path without PF's prior knowledge or consent. Through written procedure, training, and checking, most airlines have attempted to standardise these duties and ensure that changes in navigation have the consent of the Captain and the knowledge of the PF. But for a variety of reasons, these guidelines are sometimes not followed.

♦ *Among pilots operating more automated aircraft, there are comfort differences in use of automation.* People familiar with computers become comfortable more easily with automated systems. The more comfortable pilot subtly becomes a leader, and this may be contrary to chain-of-command (Wiener, 1989). That is, a First Officer who is extremely comfortable with the FMS, for example, can (but should not) make decisions best left to the Captain, and can leave the Captain out of the loop.

♦ *Among fleets of more automated aircraft, there is at least the appearance of less standardisation.* There are (objectively) more ways to complete a task on more automated aircraft (Wiener *et al*, 1990). If all of these methods are acceptable by policy or procedure, individual pilots will choose their favourite technique. As an observer looks from crew to crew and flight to flight, a great deal more techniques are observed.

♦ *Automation transfers workload among phases of flight.* On aircraft with area navigation, course changes are pre-programmed on the ground. This reduces workload during cruise and increases it preflight (Wiener, 1989). But approach information, which might have been planned in descent, is often changed in the terminal area, "requiring" on FMS aircraft, programming in addition to radio tuning for navigation, further increasing workload on approach.

♦ *Automation brings a change in timing of errors.* Beginning with INS and ONS-equipped aircraft, waypoints erroneously entered may have no effect for several hours (Wiener, 1989). A wrong course dialled into a Course Window has an immediate effect.

♦ *Among pilots operating more automated aircraft, there is a tendency to use extremes of automation.* Pilots often attempt to program when it creates new workload, or alternatively, to turn automation completely off when some intermediate mode would relieve workload. This is what pilots are taught to do when new on the aircraft - try to program, if confused, turn it off. Experience on the aircraft should lead to choosing among modes.

♦ *Pilots operating more automated aircraft sometimes have difficulty in detecting automation failures.* Pilots find it difficult to tell whether an automated function is failing or just doing it differently than he/she would do it. For example, few pilots would climb at 2000 fpm within a thousand feet of level-off. Auto level-off from the vertical speed mode on some aircraft does so routinely. It becomes possible to miss signals that capture has failed.

♦ *Among pilots operating more automated aircraft, there is a tendency to attempt to correct an "automation-induced" deviation by manipulating the automated system, rather than the controls of the aircraft.* When a deviation occurs (such as an altitude bust) while operating in an automated mode, many pilots will initially attempt to correct the deviation by manipulating the FGS or

FMS. They could instead revert to a lower level of automation, including hand flying. Reversion takes less time and may reduce the magnitude of the deviation. It is as if the pilot is competing with the machine, attempting to overcome its deviation first.

♦ *There are occasional instances of complacency or lost situation awareness when flying in an automated mode.* Crews flying coupled approaches or FMS/ONS/INS cruise flight have been unaware of a significant course or altitude deviation until alerted by ATC or another aircraft. Designers assumed that automation of flight path control and improving displays should increase situation awareness. Flight path automation should set pilots free to monitor and evaluate other things. But this goes back to the "automation as distance" issue raised above. Pilots can fall out of the loop, losing awareness of what the aircraft is doing.

♦ *Pilots occasionally suffer some loss of flying skill on return to less automated aircraft.* A number of airlines have reported that pilots transitioning from FMS aeroplanes often require additional training on return to older technology aircraft. "Feel" for the aircraft, and planning skills seem to deteriorate.

It is important to recognise that Ewell and Chidester were working from the observations of check airmen and attempting to tie those observations to research. As a result, their conclusions are more observational than systematic. They must be taken as a guide for issues to examine rather than an authoritative documentation of the human factors consequences of FMS-generation aircraft. However, where their observations are consistent with the more controlled scientific research described above, we may have greater confidence in their validity. As a result, their concerns with workload management and with mode and data awareness appear well supported. Their concerns with skill loss are more anecdotal and require further investigation.

Synthesising the issues across carriers

The ATA Subcommittee on Automation Human Factors (1998) attempted a further synthesis of these issues. This group was unique in that it involved airline operational personnel, researchers studying automation, and government personnel overseeing training and operations. In response to the FAA (1996) report, the subcommittee was tasked to define the issues into a detailed set of core problems which could lead to implementable solutions in the near term. Discussions within the subcommittee led to identification of six key issues and potential solutions that might be implemented within carriers. The policy recommendations of the subcommittee are discussed below.

Choosing among levels of automation

This issue synthesises the "automation as distance" and workload management issues defined in research. The FAA (1996) recommended that carriers provide clear and concise guidance concerning circumstances where autoflight should be engaged, disengaged, or engaged in modes with greater or lesser authority, conditions under which autoflight systems will not engage, will disengage, or revert to another mode, and appropriate combinations of automatic and manual control. The subcommittee recommended that operating manual guidance and initial and recurrent training lead to application of pilot judgement rather than provide highly specific guidance on when to use what level.

The Subcommittee noted two types of guidance provided by carriers on this issue:

♦ *Authorisation to choose levels of automation* - Each carrier has emphasised that proficiency is required in each level of automation, and has explicitly authorised pilots to choose the most appropriate level for each flight situation. Statements of this type appear today in most policy manuals, and originated with the work of Wiener (1993) and Byrnes & Black (1993).

♦ *Training for choosing among levels* - Some carriers have begun to offer guidance in training for when to choose what level, and this is a significant further step beyond simply authorising such choices. However, these training statements have not been documented in policy or operating manuals. Such guidance would include activating versus deactivating automated systems for specific/special recovery manoeuvres. For example, one carrier specified in training to disconnect autopilot and autothrottles to maintain control and extract maximum aircraft performance for: unusual attitude recovery, windshear/microburst, high altitude upset, GPWS terrain warning, engine failure at low altitude/energy state, flight instrument malfunctions, and mid-air collision avoidance.

Despite these efforts, the subcommittee noted reports to ASRS and airlines that continued to reflect two tendencies suggesting level-of-automation decision problems:

♦ *Tendency to choose an inappropriate level of automation.*

♦ *Tendency to attempt to correct an "automation-induced" deviation by manipulating the automated system,* rather than the controls of the aircraft. (FAA (1996) pointed out that in some accidents, pilots have turned to an autoflight system to correct flight attitudes beyond the capability of the system - in contrast to guidance in some airlines' training that immediate control requires manual control.)

The subcommittee recommended that more extensive policy statements be developed, consistent with recent advances in training, and that training be re-assessed to ensure pilots are being taught to operate in a manner consistent with carrier philosophy.

Detecting and correcting anomalous autoflight performance

This issue encapsulates the mode awareness problems identified in research. Pilots have reported a significant number of situations in which the aircraft deviated from the pilot's intended actions following selection of an autopilot mode. They may occur for a variety of reasons falling into two general categories - (1) inadvertent selection of a mode or input of data by the pilot, and (2) deliberate selection of a mode expecting performance different than the mode is designed to provide. The former has been referred to as input errors, and the latter as automation surprises or mode errors. In either case, the pilots must detect and correct the path of the aircraft, but the latter case reflects a deeper problem - the pilot does not understand the logic by which the mode functions in the reported situation. As a result, carriers must position their pilots to ensure that mode and data errors are quickly corrected and must position themselves to ensure that "anomalous functions" are documented in manuals and communicated in training.

The subcommittee recommended policy guidance, establishing airline reporting systems to detect these problems for documentation in operating manuals, and reporting to manufacturers to facilitate software updates and changes.

Procedural implications of functional differences in FMS and ground-based navigation

This issue grew directly from operating experience. It was not recognised by research but through pilot reports, incidents and accidents. Area navigation systems, such as FMS or GPS, navigate in fundamentally different ways from pilot tracking or autopilot coupling to ground-based navigation signals. The FMS uses inertial, satellite, and/or ground references to determine its position and navigate to waypoints defined by latitude and longitude. Pilots and earlier-generation autopilots track courses, bearings, or radials provided by a ground-based navaid. An FMS makes use of databases of waypoints, navaid locations, and procedures to accomplish the ground track for published procedures.

However, charts and databases are not identical on a variety of measures. Databases do not always correspond with charted fix names, bearings, or radials, and do not include all charted procedures. In the majority of situations, these differences are of no consequence - despite naming or depiction differences, the FMS navigates the same ground track for an approach, SID, or STAR as a pilot flying via radio navigation. However, from time to time, pilots have brought to the attention of their carriers situations where the FMS did not fly a procedure as defined by radio navigation. Reasons for differences in *approaches* were described by Transport Canada (1997) and include:

♦ Not all approaches are in the database.
♦ Waypoint identification may differ from chart to FMS.
♦ Waypoints may be added to or deleted from a procedure.
♦ Courses and bearings may differ from chart to FMS.
♦ Duplicate identifiers require database conventions that may not be apparent to the pilot.
♦ Revision cycles for charts and databases are not identical.

Pilots must determine whether they are flying the charted procedure for which they are cleared. Given the capabilities and limitations of existing navigation databases, pilots will encounter discrepancies, ranging from a simple name difference that leads to the same ground track to selection/activation or database errors that result in a substantially different course. Significant discrepancies more often result from selecting and activating a procedure incorrectly, but may also signal a database error that needs to be brought to the airline's and database provider's attention and corrected. Detection and correction requires a thorough briefing and cross-check of FMS data and charted fixes. The subcommittee recommended incorporation of these concepts into policy and training.

Display and cross-check of ground-based navaids against FMS map display

This issue also grew directly from operating experience, but its recognition owes much to the FAA (1996) discussion of unintended applications of autoflight capabilities. In general, approaches are flown relative to or from ground-based navaids. Flight Management Systems are certified for enroute and terminal navigation, but not for approaches. There are two prominent exceptions:

♦ Aircraft designed to meet Required Navigation Performance (RNP) for the Approach Phase (B-737 or B-777 with Advanced FMS, for example) may fly almost all non-precision approaches that are not based on a localiser with reference only to FMS position and guidance information. These aircraft have the hardware and software necessary to receive a general certification for approach navigation.

♦ Special FMS, GPS, and RNAV procedures may be flown on most other FMS-generation aircraft based on FMS position and guidance. These select procedures have been designed to the capabilities and limitations of FMS navigation systems.

So, a pilot may take off, accomplish a SID and its transitions, navigate enroute, and accomplish a STAR and its transitions to the initial approach fix via lateral navigation without direct reference to ground-based navaids. But by the initial approach fix, the pilot must tune, identify, and monitor the navaids that define the approach, to ensure the path flown by the aircraft complies with the required track (excepting RNP aircraft or RNAV procedures).

The subcommittee encountered a range of policy requirements to monitor ground based navaids. At one extreme, approach navaid requirements were discussed in training but not documented in operating manuals. At the other extreme, pilots were required to cross-check their position against ground based navaids throughout flight. Through these requirements, airlines were trying to prevent or correct:

♦ *Map shifts.* Occasionally, at a variety of locations (for example within the U.S. at BNA, DFW, LAX, and SAT), erroneous navaid location information has resulted in an inaccurate update to FMS position. These must be detected and corrected by the crew by reference to ground-based navaids or by ATC advisory. Detection becomes critical in the approach environment, but is potentially an issue any time the aircraft is below minimum off-route altitude (grid MORA).

♦ *Lack of position awareness while flying radar vectors in mountainous terrain.* Within the U.S., controllers routinely vector aircraft below published MSA in compliance with minimum vectoring altitudes. The level of performance by U.S. controllers has made that process acceptable, but it cannot be generalised to non-radar environments. In Latin America, for example, the threat of terrain and risk of miscommunication warrants cross-check even in a radar environment. Pilots must know where they are relative to navaids and radials to determine their safe altitude in those operations.

♦ *Approach deviations where FMS course guidance diverges from the localiser or radial defining the approach segment.* The ground track specified by the procedure must be flown, so ground-based navaids defining the procedure must be cross-checked. Any discrepancy must be resolved immediately.

♦ *Continuing an approach in instrument conditions after the failure of a ground-based navaid.* The computer-generated course line on the map display, however compelling, cannot replace the ground-based navaid defining an approach course (excepting RNP aircraft or RNAV procedures).

♦ *Terminal area deviations at certain critical airports.* Due to the frequency or criticality of deviations from SIDs or STARs as defined by ground-based navaids, some carriers have stipulated raw data monitoring and display requirements at Frankfurt and Paris, for example.

Each of these issues involves uncertainty with the location of the aircraft. The typical FMS navigation system calculates its position from an initial latitude and longitude entered by the pilot, updated by onboard inertial reference systems and distance from the known latitude and longitude of ground-based navaids automatically sampled by the FMS. At altitude, enroute, or on a departure or arrival, the accuracy of FMS navigation systems is generally satisfactory. In the approach phase, ground based navaids have provided more accurate position definition – or rather, obstacle separation has been built to accommodate navaid and receiver errors rather than area navigation errors. And some carriers have identified other situations, such as mountainous terrain, where they deem FMS position accuracy insufficient.

RNAV approaches and RNP aircraft are mechanisms for accommodating and increasing FMS navigation accuracy, respectively. RNAV approaches are built to the level of accuracy of FMS aircraft. Advanced FMS/RNP aircraft add satellite position updating to inertial and navaid information, monitor agreement among these inputs, and alert pilots when position uncertainty reaches a critical value. That is, where the mean latitudes and longitudes from inertial, satellite, and navaid reference describe a small circle, the pilot may be confident in its accuracy. When that circle grows larger, meaning the sources do not agree on position, the pilot is alerted not to rely on FMS position.

What aspects of this discussion must be understood by pilots? With the exception of RNP aircraft and RNAV approaches, cross-check of navigation against ground-based navaids is required. Pilots transitioning to and from these aircraft must understand these distinctions. Inappropriately generalising RNAV procedures to non-applicable approaches, or transferring RNP procedures to non-RNP aircraft must be prevented. The subcommittee recommended incorporation of these concepts into policy and training.

Improving ATC procedure compatibility with FMS-generation aircraft

Two issues involving ATC were raised in the FAA (1996) report or in the deliberations of the subcommittee. Last-minute runway, departure, or approach changes are a frequent source of increased workload and error on FMS aircraft. A number of ATC procedures fail to take advantage of FMS capabilities or are structured in such a way as to interfere with their use.

Errors often result when pilots attempt to comply with last-minute runway, departure, or approach changes. Part of the problem - rushing to comply - generalises to non-FMS aircraft. What is unique is an ambiguity of how to set up navigation for such a change. Should the pilot attempt to use the FMS first when a change is issued? It provides course guidance for many departures and for approaches and missed approaches along with access to the required navigation frequencies. Alternatively, since reference to ground-based navigation is required at least for approaches, should it be completed first? The subcommittee noted several reports where pilots made significant errors, including not tuning the required navaids in the rush of such a change.

Regarding ATC procedures that are not compatible with FMS aircraft, there appear to be a number of enhancements possible to existing ATC procedures that would directly benefit these aircraft without creating special procedures and without decrement to non-FMS aircraft. In general, the FMS is of little help for vector SIDs. Selection of one of the departures at New York Kennedy airport, for example may result in no segments appearing on the FMS legs page or on the Map display, depending on the runway and whether there is a common point for all climbs or transitions from that runway after takeoff. This results in a very complex departure that must be built, rather than selected, if the pilots wish to use the FMS.

The subcommittee recommended carriers provide policy guidance for handling departure, arrival, or runway changes and recommended industry-FAA collaboration on making ATC procedures compatible with FMS limitations.

Maintenance of underlying skills with extensive FMS experience

Billings (1997) has spoken of the potential of operators to become disengaged from the underlying processes that have been given over to automated systems, and described how that can result in reduced or in-ability to carry out those processes when the automated system is disengaged or disabled. To the extent that the operator must be able to carry out those processes in an abnormal or emergency situation, the underlying skills must be maintained.

Applying that line of thought to pilots, we could expect those functions that are predominantly carried out by automated systems on an FMS-generation aircraft to suffer some skill loss. This might include hand-flying the aircraft in phases of flight were the autopilot is predominantly used, thrust control when autothrottles are disengaged, calculations of climb or descent targets or timing without FMS assistance, and continuous navigation by reference only to ground based navaids.

Evidence of any change in such skills among Part 121 pilots is limited. What exists is based upon either anecdotal reports by Training departments (Ewell & Chidester, 1994) or upon accident, incident, or event reporting and analysis (FAA, 1996). Airlines have described possible examples including additional training required by First Officers with extensive FMS experience upon upgrading to Captain on less-automated aircraft (instrument scan and descent planning are frequently cited), activation - rather than deactivation - of autoflight systems in recovering from excursions from controlled flight in at least two accidents, and events in which pilots have assumed disengaged autothrottles were engaged and applied no thrust control as the aircraft accelerated or decelerated out of the desired flight envelope. It is clear that many airlines have concluded that maintaining underlying skills is a challenge, with or without controlled scientific studies to back those beliefs.

But perhaps the best evidence for skill loss with extended use of autoflight and flight guidance/management systems has come from the recent decertification of Omega Navigation Systems. A surge in navigation deviations on aircraft formerly equipped with ONS and not yet equipped with GPS was observed in reports to airlines and ASRS in late 1997. The source of these deviations was typically a failure of the pilot flying to tune and identify a new navaid, or select a new or correct radial on station passage, resulting in failure to make a required turn. Similar errors involving incorrect calculation of segment distances resulted in early course turns. These are basic functions of instrument navigation that were, until recently, assumed by ONS coupled to the autopilot. When that function was removed, pilots had difficulty in reapplying a well-learned and understood process they had performed throughout their careers. This is exactly the phenomenon Billings describes in more general terms.

Summary

The synthesis of issues by the ATA subcommittee can be viewed as an industry agenda for the issues to be addressed by pilots and airlines. It pulls together both what has been identified in research and nearly two decades of operating experience. But the reader should also recognise what the subcommittee report did not do - it did not set the agenda for design issues in the next generation of displays and controls. That some of these issues ultimately require design or certification solutions is recognised in the FAA (1996) report, which dedicated many of its recommendations to revising the certification process. Both approaches - dealing with the current generation and building better future designs - are necessary. The ATA subcommittee recognised that action was required in the near term by airlines or their pilots to prevent and correct commonly occurring errors. That is the focus of this chapter, and I will use the issues from that report as an organisational framework for the following section.

Developing solutions at the operator level

Responding to the human factors issues identified in FMS-generation aircraft will take actions within airlines. Aircraft and pilots are immersed almost completely within the context of an individual airline. If airlines are to bring about the pilot actions necessary to correct the trends discussed above, they must look to policy, procedures, and line practices (Degani & Wiener, 1991). These are the vehicles by which they communicate and standardise their expectations for pilot performance. Training and evaluation most directly affect line practices. The ATA Subcommittee (1998, 1997) made recommendations for improving both policy and training.

Developing policy

Each of the human factors issues discussed above suggests a knowledge or policy gap. The subcommittee drafted policy guidance carriers might publish to their pilots on the six issues they identified.

Choosing among levels of automation

Application of pilot judgement is the immediate target of guidance on this issue. The subcommittee offered the following prototypical statement for publication in policy manuals:

> Pilots will be proficient in operating their aircraft in all levels of automation. However, the level of automation used at any time should be the most appropriate to enhance safety, passenger comfort, schedule, and economy. Pilots are specifically authorised to choose what they believe to be an appropriate level of automation. In

general, choices among levels can be guided by their functionality and the demands of the situation.

Where immediate, decisive, and correct control of aircraft path is required, the lowest level of automation - hand-flying without flight director guidance - may be necessary. Such instances would include escape or avoidance manoeuvres (excepting aircraft with flight-director windshear guidance) and recovery from upset or unusual attitudes. With the exception of visual approaches and deliberate decisions to maintain flying proficiency, this is essentially a non-normal operation for flight guidance or FMS-generation aircraft. It should be considered a transitory mode used when the pilot perceives the aircraft is not responding to urgent aircraft demands. The pilot can establish a higher level of automation as soon as conditions permit.

When used with flight director guidance, hand flying is the primary takeoff and departure mode. It is also the primary mode for landings, except for autolands.

Where short-range tactical planning is needed (i.e., radar vectors for separation or course intercept, short-range speed or climb rate control, etc.), Mode Control or Flight Guidance inputs may be most effective. This level should be used predominantly in the terminal environment when responding to clearance changes and restrictions, including in-close approach/runway changes.

Autoflight coupled to the FMS/GPS is the primary mode for non-terminal operations and should be established as soon as "resume own navigation" or similar clearance is received. This level exploits programming accomplished pre-flight. Where the longer-range strategic plan is changed (i.e., initial approach and runway assignment, direct clearances, etc), Flight Management inputs remain appropriate. However, when significant modifications to route are issued by ATC, the pilot should revert, at least temporarily, to lower levels of automation.

These recommendations represent an expansion of the automation policy recommended by Byrnes & Black (1993), which authorised judgement but did not recommend criteria for applying levels to each situation.

Detecting and correcting anomalous autoflight performance

Events of this type represent underlying misunderstandings of the functioning of particular modes by the pilot, a lack of strategy for confirming modes engaged or annunciated, or unusual or counter-intuitive entries into a mode or envelope. Alternatively, they may be viewed as inaccurate design assumptions of how a pilot would use or encounter a mode, or how an ATC clearance would affect mode selection. Countermeasures are necessary at three levels:

♦ Pilot strategy – pilots must deliberately scan the FMA to determine whether autopilot and autothrottles are engaged and in what modes. Failures of mode awareness often reflect a failure to continue scanning the performance of the aircraft following selection and confirmation of an autoflight mode.

♦ Airline reporting – where seemingly anomalous autoflight performance can be traced to a design assumption or software, it must be documented to pilots in operating manuals and reviewed in training. Many of these events reflect a mode functionality expectation by the pilot that is not valid. If these can be documented and/or highlighted in training, they may be overcome by changing the expectation. This requires capturing the events though a reporting program.

♦ Manufacturer reporting – where large numbers of similar events are found to be associated with a particular design assumption or software implementation, software updates and changes are warranted. Each manufacturer has implemented such changes through software upgrades. They can accomplish such corrective actions only to the extent they are made aware of events by their airline customers.

The ATA subcommittee recommended that airlines review their procedures and training to assess the extent to which they promote a pilot strategy of autoflight use that confirms the annunciation of selected modes and continues scanning for anomalous performance:

Pilots must confirm the results of autoflight selections to prevent mode or course surprises and confusion. A selection on the Mode Control or Flight Guidance panel must be checked against its result on the Flight Mode Annunciator. Input into the FMS/GFMS-CDU must be checked against its resulting course displayed on the Nav Display, and *the pilot making the input must confirm the resulting course with the other pilot prior to*

executing the change when feasible. And in all cases, both pilots must continue their scan to ensure the autopilot performs as directed and anticipated.

Airlines will also need to pursue pilot reporting systems to capture these events and report them on the manufacturer where appropriate.

Procedural implications of functional differences in FMS and ground-based navigation

The ATA subcommittee recommended that airlines review the extent to which their pilots need to know the underlying database issues, but more importantly, the extent to which their procedures emphasise the pilot actions required to ensure the aircraft flies the track required by a procedure:

> For a variety of reasons, displayed FMS legs making up a departure, arrival, or approach procedure may not correspond with charted fix names, bearings, or radials even though the database is designed to follow the same ground track. However, from time to time, pilots have encountered situations where the FMS did not fly a procedure as defined by radio navigation or in compliance with ATC expectations. Therefore, pilots must brief and cross-check charted procedures against FMS data to ensure they have selected the correct procedure and will comply with their clearance.

♦ Before departure, thoroughly review your assigned departure and cross-check the waypoints obtained with your desired course. If you select or build a transition, verify between pilots that it matches your clearance and produces the desired track. Ask ATC for clarification if any conflict exists.

♦ Before arriving in the terminal area, thoroughly brief the arrival and approach you expect to fly and cross-check fixes presented by the FMS against fixes depicted on the approach chart. Should the runway or approach change and you wish to use the FMS for the new approach, that same level of cross-check is essential. If time constraints or circumstances prevent your cross-check, decline the clearance or tune and identify radio aids to navigation and fly the approach in a lower level of automation.

Display and cross-check of ground-based navaids against FMS map display

The ATA subcommittee recommended that airlines evaluate and define their requirements for reference to ground-based navaids:

Except for those aircraft designed to meet Required Navigation Performance (RNP) for the Approach Phase (B-737 or B-777 with Advanced FMS, for example), Flight Management Systems are certified for enroute and terminal navigation, but not for approaches. Except where prohibited by bulletin or company-specific pages in the Airway Manual, pilots may accomplish a SID and its transitions, navigate enroute, and accomplish a STAR and its transitions to the initial approach fix solely by FMS navigation, *but not approaches*.

Except for published FMS, GPS, and RNAV instrument approach procedures, approaches are flown relative to ground-based navaids. For all other approaches, prior to the initial approach fix, one pilot must tune, identify, and monitor (on a CDI display, where available) the navaids that define the approach. These actions are necessary to ensure the path flown by the aircraft complies with the ground track required by the approach procedure. The function of the FMS and Nav display during an approach is to assist your situation awareness - not to fly the approach. Any discrepancy between the Nav Display or Flight Director based on FMS/GFMS guidance and raw data from navaids defining the approach must be challenged and resolved immediately. Should the ground based signal be lost, the crew must abandon that approach if in instrument conditions. On all instrument approaches inside the final approach fix in IMC weather conditions, a go-around is required whenever unreliability or full scale deflection of the ground-based approach navaids is encountered.

Improving ATC procedure compatibility with FMS-generation aircraft

The ATA subcommittee also concluded that further guidance to pilots on response to ATC clearance changes in the terminal area is warranted:

Proper use of automation will reduce your workload, freeing you to complete other tasks. Improper use will do just the opposite. Whenever possible, avoid FMS/GFMS programming during critical phases of flight. Complete as much programming as possible during low workload phases. ATC clearance changes in the terminal area directly challenge this requirement.

A departure change during taxi for takeoff requires review of the assigned departure. If the FMS is to be used for navigation during the departure, pilots must cross-check the waypoints obtained with the desired course. However, pilots may choose to navigate the departure by ground-based navaids if update and cross-check of FMS moving map displays would distract from primary ground and flight duties.

While pilots must tune, identify, and monitor all applicable approach navaids for every approach and landing, it is not necessary to update FMS moving map displays close-in to the landing airport where "heads down" data entry would distract from primary flight duties.

Maintenance of underlying skills with extensive FMS experience

Though conceding the evidence is mixed, the ATA subcommittee was concerned about two issues - maintaining skills that remain necessary in FMS aircraft, and skills that are necessary when transitioning to a less-automated aircraft. They defined the following as potential concerns:

♦ Assembly of situation awareness from disparate instruments rather than only from a map display remains a critical skill, but can become unexercised operating an FMS aircraft in a non-threatening environment. Map displays present valuable and key elements of situation awareness, but not all of them. For example, until the advent of EGPWS, no terrain information was displayed. Pilots have always been required to assemble position, course, and terrain information in some form of mental map described as situation awareness. With map displays, some but not all elements are automatically displayed or are pilot selectable. Do pilots routinely practice integrating non-displayed elements in non-threatening environments? This seems unlikely, and if not, represents a challenge to maintaining these skills. Such skill is occasionally critical, as in the case of a map shift, or in recovery from an inadvertent track deviation (Woods & Sarter, 1998).

♦ Instrument cross-check condenses toward the primary flight display on FMS-generation aircraft; more so with integrated primary flight displays. Cross-check of other instruments is necessary in certain phases of flight and this requirement broadens with certain types of approaches.

♦ Similarly, with the known high reliability of FMS navigation, PNF monitoring skills may go unchallenged – if deviations are very rarely detected, the motivation to search for them naturally declines. But, the PNF must be as alert as the PF, regardless of level of automation available in the aircraft. A low probability, high criticality error is exactly the one that must be caught and corrected.

♦ Flight path management, though a key function of FMS, continues to require pilot judgement skill. For example, while calculating descents is automatic, the pilot must do so quickly in response to ATC-imposed crossing restrictions. This remains necessary because the validity of these clearances must often be assessed in less time than is required to set them up in the FMS. Otherwise, the pilot discovers that the opportunity to make the restriction passed while he/she and the FMS were calculating it.

The Subcommittee recommended that airlines look for opportunities to test and reinforce these skills in training and checking.

Developing training

With a sense of the key issues on FMS-generation aircraft and a set of policies to deal with them, airlines are re-examining how they train pilots to fly these aircraft. The ATA Subcommittee (1997) offered a critique of the first generation of training programs for FMS aircraft and suggested some alternatives for redesigning training to make it consistent with operating policy.

Inconsistency of first-generation training programs with operating philosophy

Most qualification training programs were not designed to be consistent with operating philosophy statements. Earl Wiener, participating in the deliberations of the subcommittee, pointed out that the initial training programs for FMS aircraft were essentially adaptations or minor re-writes of training for previous generations of aircraft and resulted in failure rates approaching 20%. This led almost immediately to revised approaches unique to FMS aircraft - the "first generation". Because pilots who did not succeed had difficulty mastering FMS and higher levels of automation, revisions were designed to ensure mastery of flight management, biasing training time and focus toward that level. Checking likewise became biased towards what was perceived to be most difficult on the aircraft. Both trends were to the detriment of operating at lower levels and demonstrating judgement. Most of these first-generation automated aircraft training programs remained in service until very recently.

If we accept carrier statements authorising pilots to choose among levels of automation (e.g. Byrnes & Black, 1993), the principal flying tasks of a pilot operating these aircraft are competency in hand-flying, mode control, and flight management levels and judging how and when to apply each level. These represent at least two new skill areas for pilots transitioning from B-727 generation aircraft and one new skill area for pilots transitioning from MD-80 generation aircraft. Yet, the qualification training cycles for all three generations are very similar in length across fleets and carriers. It is unlikely that all three skills can be *taught* in this same timeframe, offering some explanation for both the biases toward the new and higher levels, and the FAA's (1996) concerns about level of knowledge and application of judgement.

Wiener argued to the subcommittee that checking at the highest levels, and training to the checkride, will miss problems at lower levels, will not check the most important criterion - judgement - and will reinforce beliefs that highest levels of automation are the most proper for line operations. Training failure rates have fallen dramatically, but the focus on higher levels has produced a mindset contrary to carrier operating goals and consistent with the problems identified by the FAA Team. *In a sense, we have taught the test, rather than our operational goals.*

Several carriers have come to similar conclusions in their own self-assessments. Assuming competency and de-emphasising training events at the lower levels in favour of training and checking at the higher levels of automation may result in pilots biasing their skills and choices to the highest levels. From the pilot's perspective, how we train to fly these aeroplanes is centred on the FMS - so flying becomes centred on the FMS.

Approaches toward designing training consistent with philosophy

If first generation programs were inconsistent with carrier operational philosophy, how would training be designed to teach the way the carrier wants the aircraft flown? At least three approaches have been proposed, training the basic aircraft first, revising training by ongoing feedback, and most recently, training all levels and judgement.

♦ Training the basic aircraft first

A case can be made for training the basic aircraft first. In this approach, pilots would attend a day of ground school overviewing systems unique to an FMS-generation aircraft and explaining the airline's operating philosophy. However, further discussion of autoflight systems would be deliberately deferred for some number of days of ground school or device training, so that other systems and basic flight functions can be emphasised. The underlying philosophy is consistent with those described above - if we expect pilots to revert to more basic levels or modes of operation, we must train them, rather than assume proficiency at those levels. What remains unresolved is whether this meets a second critical concern raised by the FAA (1996) - ensuring pilot knowledge and proficiency in all modes and functions. It seems unlikely that pilots can absorb this full range of knowledge in a foreshortened subcourse when this is their area of least experience. First-generation programs were biased toward higher levels as a consequence of pilots failing to readily absorb these new skills. As a result, the subcommittee viewed this approach as potentially trading *pilot proficiency across levels* for *proficiency at intervention* at lower levels. While the *consequences* of mode confusion might be reduced, the *incidence* might actually increase. Is it necessary to trade one set of problems for another, or can both goals be achieved in comparable-length courses?

◆   Revision by ongoing feedback

A case can also be made for building upon existing training technology through feedback from training, line observations and reported events. This is the primary emphasis of Advanced Qualification Programs implemented by the FAA and participating carriers under special FAR 58. For example, one airline received comments from crewmembers completing IOE that they found visual approaches to be the most challenging task on the line, and they did not believe training for visuals was adequate. The carrier responded by building several visual approach manoeuvres into the training program.

With the recent development of partnership programs (where pilots report events under and agreement among their airline, pilot union, and the FAA) and Flight Operations Quality Assurance (FOQA - where airlines monitor flight data recorders for flight anomalies under and agreement with their union and the FAA) it appears that feedback should increase in quantity and quality. Recurrent training data collected under AQP can be analysed or restructured to provide such feedback. Two key tests await such an approach - the extent to which data can be gathered to ensure valid and systematic, rather than anecdotal, information and the extent to which pilot performance becomes congruent with the company's operational philosophy. The subcommittee concluded that feedback will be a critical element in any approach, but was uncertain it would be adequate to accomplish the redesign necessary to address the concerns and recommendations of the FAA (1996).

◆   Training all levels and judgement

Most airlines are moving toward this approach, which attempts to ensure that all levels of automation are presented and checked within a qualification program and that judgement tasks are represented within training and checking. The subcommittee selected one such program to describe this approach in some detail. Its designers deliberately chose not to assume proficiency in lower levels. Instead, they chose to train the FMS and autoflight systems as they would any other system on the aircraft - determining which aspects are best taught in ground school, which in system trainers, and which in flight simulation. Knowledge and skills associated with autoflight systems were examined to determine where they could best be taught and checked. This allowed distributing different aspects of training among ground school, FMC trainers, and flight simulators. It can be described as spanning five phases:

♦ *Overview and operational philosophy.* The first phase of automation training is included in indoctrination training and is provided the first time a pilot qualifies on an FMS-equipped aircraft. The course presents the airline's philosophy on use of automation, as well as a basic overview of cockpit instrumentation, displays, symbols, FMS, etc. The training focuses on what makes up an automated cockpit, the sources of information available, what to expect if one or more sources are lost, and why the automation systems are designed the way they are. The concepts of "managing" automation and various "levels" of automation also are introduced.

♦ *Ground training of FMS pages.* The second phase is integrated into the ground training segment and consists of basic overviews of the specific aircraft's flight management system, its controls and indicators, CDU page format and design layout, descriptions of the functionality of each page, and system limitations and abnormalities. This phase lays the knowledge-based foundation that will be required to later use the FMS in an operational context. FMS pages are introduced each day during ground training in self-study lessons, followed by training with a ground school instructor on a flight management system part-task trainer (FMST). The process repeats itself each day during ground training until all pages are covered. The pilot's knowledge of these basic FMS pages is tested during the oral evaluation.

♦ *FMS procedures.* The third phase is conducted as the last module in the ground segment, and is taught by a simulator instructor using the FMST. This module builds upon the basics learned in the previous phase by teaching FMS procedures that are used in the simulator and on the line. Typical FMS procedures covered are initialisation, flight plans, takeoff, departure/enroute, approach, and miscellaneous advanced functions. The objective of phase three is to ensure the pilot understands and can properly use the FMS during most flight operations without the distractions of other aircraft systems and having to "fly" the aircraft. Before the pilot can progress beyond this module in ground training to the flight training segment, the instructor must verify proficiency in FMS procedures.

♦ *Manoeuvres at each applicable level of automation.* The fourth phase is conducted during the simulator training segment. This phase integrates the previously learned FMS procedures in flight simulator scenarios. It is important to note that the emphasis in this automation phase is on manoeuvre proficiency at various levels of automation. Literally, each required manoeuvre is demonstrated in each applicable level of automation. The level of automation to be used on each event is dictated by the syllabus. In addition, the same computer-based FMST used in ground training is available for use in the simulator briefing rooms. During the pre-brief prior to simulator training, the instructors and pilots can set up and discuss various FMS scenarios, then upload the initialisation and scenario data to the simulator. During the debrief, any FMS problems that occurred during simulator training can be reconstructed for effective learning and correction of errors.

♦ *Applying judgement to problems and scenarios.* The final phase is conducted in the last few simulator periods. This phase emphasises applying judgement in the proper use of automation. The instructor selects the problem. The "level" of automation selected in any event is at the discretion of the pilot. And the instructor provides feedback on that decision. As described above, the computer-based FMST is available in the simulator briefing rooms as an aid in learning. The subcommittee recommended this type of approach to training for FMS-generation aircraft. Training should be structured to be consistent with operating policy then revised by ongoing feedback.

LOFT for FMC aircraft

Line Oriented Flight Training (LOFT) is a technique by which a training period is conducted in the form of a line flight rather than a series of selected manoeuvres. The goal of LOFT is to provide crews with a challenging problem to resolve within the constraints of a line operation. Done well, this format brings together the workload and operational challenges often observed in accident scenarios, and assists the crew in developing their technical proficiency, situation awareness, problem-solving, and co-ordination skills. This would appear a necessary component to training for FMS-generation aircraft. However, LOFT for these aircraft has often differed little from scenarios on previous-generation aircraft.

I believe we have reached a point where airlines can better implement LOFT on FMS aircraft. To accomplish that, training managers may incorporate each of the issues discussed in this chapter into LOFT scenario design. For example, one carrier decided to build its LOFT scenarios around challenges to situation awareness. The training managers decided that all fleets would use airports in the vicinity of mountainous terrain in their LOFTs and incorporate failures that constrained climb or high altitude performance, made a diversion decision necessary, or provided some navigation system problem. Their objectives across fleets, then, were to improve management of situation awareness in the context of high workload while operating in a threatening environment. An added benefit to technical proficiency coincided, as the airline had recently introduced an improved terrain clearance program on all flight plans. This could be reviewed and applied in a line-oriented training setting.

The FMS-aircraft training managers could have stopped there and provided a valuable training program. But they chose to select airports with potential navigation database anomalies or challenges as a substitute for navigation systems failures, tailoring LOFT to FMS challenges. For example, one scenario departed Guatemala City, encountered one of several options for systems failures which would cause the crew to return. In their descent, the crew received clearance for an approach that was not in the database. This provides an opportunity to build knowledge on several of the issues identified on FMS-generation aircraft. Pilots flying this scenario will have to make a levels-of-automation decision. If using flight management, they will have to carefully crosscheck their approach waypoints against the chart to recognise an incorrect approach selection unless they understand the database limitations in advance. And they must maintain their position awareness while dealing with this anomaly and the diversion.

Consider then, some options for introducing the human factors issues on FMS-generation aircraft into LOFT. To challenge level-of-automation decision-making, design scenarios which thwart an existing plan of action - introduce strategic change, such as diverts in proximity to the landing airport. These scenarios tend to concentrate workload and force these choices. To challenge detection and correction of anomalous autoflight performance, choose routings and clearances that play to known issues on the particular aircraft - for example, departures with fly-over waypoints and route discontinuities, or common mistakes reported by pilots or check airmen. To challenge understanding of FMS navigation of charted procedures, include departures, arrivals, and approaches that are not databased at all, or are not completely databased. Better still, introduce these as a change in plan on departure or during a diversion. To challenge cross-checking of charts and ground-based navaids, select airports where not all approaches are in the database, such as the Guatemala City LOFT above. To challenge the maintenance of underlying skills, implement failures with consequences for some or all levels of automation - examples can range from failure of an autopilot, FMC, or flight director, to an underlying system used by all these systems such as pitot-static system failures. This approach takes advantage of the line constraints represented in all LOFTs, but tailors them to the unique technical and human factors of this generation of aircraft.

Summary

In summary, training for FMS-generation aircraft typically preceded the development of operating philosophies, and as a result tended to bias toward the flight management level of automation. Airlines have begun in the past few years to conduct thorough reassessments of the form and content of their training for these aircraft, with a consensus building around strategies that deliberately train at all levels and present opportunities to apply and receive feedback in judgement. LOFT for FMS-generation aircraft represents a special case of tailoring training, with great potential for improving performance.

Conclusion - lessons for the next generation of aircraft

Integrating FMS-generation aircraft into line operations represents a nearly two-decade struggle to implement new technology into an operating environment. What does this experience tell us about integrating future technology? Our industry has learned a number of lessons. While much of the discussion of this chapter is unique to this generation of aircraft, the fact of unexpected human factors consequences of new technology is not. Design of new technology requires a vision of both present and future environments in which technology will operate. The introduction of new flight technology requires assessment of policy and procedure within the airlines that will operate them. Look to some of our assumptions and their outcomes for FMS-generation aircraft:

♦ Airlines tended to assume proficiency at lower levels of automation among their pilots and biased their training time and events to the new, higher levels introduced in FMS-generation aircraft. Their pilots often observed, learned, and applied a categorically different way of flying this new generation of aircraft (ATA, 1997).

♦ Designers assumed that pilots encountering anomalous autoflight performance would disconnect the offending system. Many reported events, some incidents, and a few accidents reflect instead a struggle to make the system perform as desired, or overcome an anomaly the pilot did not understand or recognise (FAA, 1996).

♦ Workload was a central issue in the design of the current generation of aircraft. Error tolerance and error correction have been found to be equally important, while the relationship between automation level and workload has been found complex (Billings, 1997; FAA, 1996; Wiener *et al*, 1990).

We have seen policy grow from recognition of problems and recognition of assumptions that turned out to be invalid within the *operating context*. I doubt these were completely predictable in the timeframe of design of FMS-generation aircraft. I would suggest instead that issues in the interaction of new technology with the existing operating environment are not and perhaps cannot be anticipated entirely at the design level, even where human factors specialists are a part of the design process. Airmen will apply novel, unintended uses of technology, which may exceed design expectations and limitations, or alternatively, reflect improper premature application of future capabilities. We must act, then, at several levels to safely introduce new generations of technology in airline operations.

At the design level, the FAA (1996) has emphasised operational or performance testing of new systems early and often in the process and incorporating this philosophy into certification criteria. This makes sense, but I would also offer a caution. Designers are often presented with the double-bind of making a system work well within the present airspace system, and making it enable a future navigation system. For example, those who developed area navigation capabilities would not have served us well by perfectly emulating today's ground-based navigation system, but making RNAV procedures unfeasible. Conducting performance testing with subjects grounded in the present situation is necessary, but not sufficient. It provides feedback on the ease of use of a design concept among those who will operate it, but not necessarily in the context in which it will ultimately be applied. Both views are necessary.

At the airspace system level, we need to enable the use of more capable systems sooner. Making FMS-generation aircraft perform like the previous generation exacerbates many of the issues discussed in this chapter. As these aircraft have approached and exceeded a majority of air transport aircraft, more FMS procedures have been developed. This should become a major focus of work within the FAA and JAA air traffic divisions. As noted by the ATA Subcommittee (1998) many departure procedures are designed incompatibly with the underlying structure of FMS databases. By withholding new capabilities, or holding them to current procedures, we may enable or encourage error. This warrants more focus now and with the next generation of aircraft.

Two different types of actions are warranted at the airline level. As an airline purchases new aircraft, devices, or systems, a careful review of its policies, procedures, and training in light of new capabilities is necessary. A seemingly innocuous upgrade that brings with it substantial new capabilities will require policy change to enable the benefit and prevent rational, but unintended uses of the capability. Without operator examination, policy and training for new technology will not capture that set of issues. Further, following introduction, operators must position themselves to identify and correct anomalous functions or uses in the line environment. Operators will discover unexpected issues through line feedback systems. The speed of revision of policy, procedure, and training will greatly depend on the quality of the feedback system.

In closing, I would point out that these challenges are being presented right now, in the form of recently implemented Enhanced Ground Proximity Warning Systems (EGPWS) and RNP approach capabilities. Both of these systems could be misapplied - EGPWS by navigating laterally on the basis of the visual image presented by selection or warning activation, RNP by flying a non-applicable approach by RNP procedures or generalising them to non-RNP aircraft. It is my observation both from within an airline and as a member of inter-airline working groups that airlines are spotting these potential misapplications earlier, implementing policy guidance along with the systems, and publishing reported anomalies to the pilots quickly when they are observed. This would suggest we have demonstrated applied learning - recognising the human factors consequences of new technology and doing our best to correct them. If that is the legacy of this fundamental shift that Billings (1997) christened FMS-generation aircraft, then our industry and our passengers will be well-served.

Acknowledgements

The ideas expressed in this chapter were greatly influenced by my fellow members of the Air Transport Association Subcommittee on Automation Human Factors. Particularly worth recognition are the contributions of Frank Tullo, Bruce Tesmer, Peter Wolfe, Corky Romeo, Jesse Marker, and Steve Predmore who served throughout the committee's tenure. Without their commitment and contributions, this chapter could not have been completed. The Subcommittee also owes much to the human factors experts who advised us, including Charles Billings, Earl Wiener, Kathy Abbott, and Eleana Edens. The sponsorship of the Air Transport Association and its member carriers enabled the collaboration reflected in the subcommittee reports, which formed the intellectual foundation of this chapter.

11 Automation and Advanced Crew Resource Management

THOMAS SEAMSTER

Cognitive and Human Factors, USA

Flight deck automation training has been developed as a set of independent modules, with general automation practice provided to crews during Line Oriented Flight Training (LOFT). Similarly, airlines have implemented Crew Resource Management (CRM) training as modules introducing general principles to improve crew performance and crews are provided with general practice in LOFT sessions. The result has been that both automation and CRM have been trained more as additions to, rather than as a systematic part of, Standard Operating Procedure (SOP). From an historical perspective, this independent, modular approach to training is understandable, but with increasing training demands, especially in automated fleets, airlines need greater integration between automation and CRM and also between those two elements and flight operations.

Programs such as the Advanced Qualification Program (AQP) have provided fleets with the opportunity to develop a tighter integration intra automation and crew resource management as well as inter operations. The intra automation and CRM integration should be based on the training of specific skills, something that airlines have started doing, at least on the CRM side. Integration inter, or across operations, should be based on a consistent philosophy, policy, and set of procedures that support individual and crew skill development.

This chapter presents a framework for training human and computer resource management skills and procedures in the context of aviation operations and the organisation. This implementation framework considers automation and management skills and procedures from the individual pilot, from the crew, and from the organisation perspectives. The framework is partially based on the analysis of CRM skills which has identified common or compatible skills between automation and CRM, especially at the crew level.

It is further based on the implementation of CRM procedures within SOP. Although the examples are from US airlines, the implementation framework can be adapted to most carriers to help them integrate automation and resource management with flight operations.

Introduction

Flight deck automation has grown in commercial fleets changing the nature of pilot tasks. Many of the changes have improved overall pilot performance and system safety, but there are a number of "vulnerabilities" (FAA, 1996) that require the attention of the aviation community. These vulnerabilities are caused by the interaction of the design, implementation, regulation, and training of automation with the people and processes in the aviation system. The approach to evaluation, the criteria, and methods used in design, certification, implementation, and training are part of the complex interaction that has left the system open to vulnerabilities such as the pilot's understanding of the automation and its modes and judgement on when to use different levels of automation. Resolution will require an integrated effort across designers, regulators, and operators. This chapter concentrates on the factors under the direct control of the operators and presents an implementation framework to be used primarily by the airlines in training and implementing automation. This is not to say that the designer and regulator components are not significant, rather it is to focus attention on factors that airlines can address immediately as design and certification work on the factors under their control.

Automation training, like CRM training, has often been implemented as a set of additional modules that must be mastered by pilots along with the other aircraft systems. Even though airlines use LOFT sessions to promote integrated training, both automation and CRM are still viewed more as additions to, rather than as part of SOP. The first part of this section introduces flight deck automation and then presents an overview of CRM training to show some of the reasons it has been difficult to integrate CRM with flight operations. Next, a program is introduced which has recently succeeded in developing a set of CRM procedures. This program involved working with the key levels of an airline's organisation to better integrate CRM with operations. This type of program could be expanded to include automation providing carriers with a process for a more comprehensive integration.

Flight deck automation

Flight deck automation is the execution of human pilot activities by computer systems. As Parasuraman and Riley (1997) point out, this is a moving target because once specific activities cease to be performed altogether by humans, those activities are considered machine functions and are no longer treated as automation. For this paper, flight deck automation is exemplified by the Flight Management System (FMS) found in many, but not all glass cockpits.

Automation, when limited to the FMS and related systems such as the Mode Control Panel (MCP) and the Computer Display Unit (CDU), is a relatively advanced, but not overly complex system. As the pilots interact with the system under sometimes difficult operational conditions, then flight deck automation becomes a complex system with a myriad factors affecting its use. From the point of view of the pilot, there are a number of factors that determine how and when flight deck automation is used (Parasuraman & Riley, 1997). One set of factors are related to the degree of confidence, reliance, and trust that pilots have in the automation. Other factors are related to the skill and accuracy of the operator. Additional factors include the task at hand, the perceived pilot workload, and level of fatigue experienced by pilots.

In considering the entire aviation system from the perspective of airlines, manufacturers, and regulators, there are superordinate factors influencing how automation is used (FAA, 1996). One set of factors are related to the communication between key organisational elements. There are also factors connected to human performance issues associated with automation design. Further, there is the critical automation knowledge and experience throughout the aviation system. Finally, at the international level, there are cultural differences affecting automation and its use. These pilot and system wide factors can combine to make the use of flight deck automation highly complex and dynamic.

This brief overview underscores the complexity of flight deck automation as it interacts with the aviation system. This chapter does not address all the factors, rather, it concentrates on the factors within the control of airlines. From an operational perspective, airlines need an implementation framework that will allow them to manage the key factors affecting automation use that within their immediate or short term control. These include factors related to policy of how automation is used, procedures that specify automation actions, and crew automation training.

Crew resource management training

Airlines have implemented CRM training with an emphasis on concepts and principles that improve crew performance and flight safety. With up to 20 years of experience in training CRM, airlines have taken a range of approaches to its content and structure. Across these varied approaches, there are some trends that have resulted in CRM at most airlines being trained and assessed as additions to, rather than as part of SOP.

Airline CRM training started about 1980 and has undergone substantial change since its inception (Helmreich, Merritt & Wilhelm, 1999). Early CRM training, when it stood for "Cockpit Resource Management", concentrated on psychological concepts related to interpersonal behaviour. A substantial change came in the mid 1980s when the term, "Crew", replaced "Cockpit", and the training became more relevant to aviation with an emphasis on team building and on briefing and stress management strategies. During the first ten years much of CRM training was viewed by pilots as psychologically based in part because of its emphasis on their attitudes (Gregorich, Helmreich, and Wilhelm, 1990) and on their interpersonal and managerial styles. Training during this period was relatively modular, emphasised concepts rather than skills, and used non-operational classroom simulations and games.

In the early 1990s, initiatives such as the Advanced Qualification Program (FAA, 1991) in the US encouraged airlines to more fully integrate CRM with technical training. As airlines moved beyond issues of attitude and managerial styles, they emphasised behavioural markers and management skills. These markers and skills were clustered around concepts of crew co-ordination, decision making, situation awareness, and workload management. The markers were often presented as recommended practices with crews encouraged to implement them when and how they saw fit. Unfortunately, the resulting behaviour was not always predictable, and most airlines have found it difficult to specify standards of performance for CRM markers and principles.

Emerging approaches include the development of CRM procedures and a move to shift from crew resource management to an emphasis on error management (Helmreich, Merritt & Wilhelm, 1999). As with many changes over the past 20 years, these current emphases are just one piece of the needed solution. They provide either a method, a focus, or a product, but they do not provide an overall process that airlines can use to integrate and then continuously improve crew performance. The next subsection introduces an approach that has been successful at integrating CRM with SOP.

Integrating crew resource management with procedures

One approach to integrating CRM with flight operations has been demonstrated through Advanced Crew Resource Management (ACRM). Capitalising on the emphasis of CRM skills and working with US airlines restructuring training under AQP, ACRM incorporates CRM practices into normal and emergency SOP (Seamster, Boehm-Davis, Holt & Schultz, 1998).

ACRM presents a complete process to help airlines to transition CRM elements into airline-specific normal and non-normal procedures to benefit crew performance.

The ACRM process includes identifying and developing the CRM procedures, training of the instructors and evaluators, training of the crews, and a uniform assessment of crew performance with an ongoing implementation framework. ACRM has been designed to provide airlines with unique CRM solutions tailored to their operational demands. Procedures are developed to emphasise CRM elements by incorporating them where they most benefit normal as well as abnormal and emergency SOP. A key to ACRM success distinguishing it from most existing CRM training programs, is that it is an ongoing, dynamic, development process and not just new procedures, modified training, or improved operating documents.

It is important to note that reproducing a briefing card or imposing a set of CRM procedures will not, by itself, produce the type of organisational change that is essential to achieving long-lasting CRM improvements. Airlines are open to importing successful practices from other airlines, and they have been know to borrow or copy the training or operational products without adopting the underlying structural changes. This "shortcut" often results in less than satisfactory long-terms changes. As has been pointed out in a recent review of the evolution of CRM, most forms of CRM training fail to reach all crews and key CRM concepts can decay within the pilot community over time (Helmreich *et al.*, 1999). ACRM encourages a generative process, and not just the products, to reach all crews and to keep its content up to date by addressing current operational and training issues.

ACRM is directed to the training and assessment of CRM skills with the CRM procedures becoming a focal point in that training allowing crews to practice specific behaviours both in normal and non-normal situations. The procedures help crewmembers develop a consistent pattern of crew co-ordination allowing crews to know what to expect from each other. These procedures also serve as a constant reminder to the importance of CRM within the operational environment. As an integral part of SOP, CRM procedures may be integrated within briefings, checklists, and emergency or abnormal procedures, such as those found in a Quick Reference Handbook (QRH), the Flight Standards Manual (FSM), or the Flight Operations Manual (FOM). For example, crew communication and situation awareness can be improved by requiring specific items in briefings prior to takeoff. A takeoff brief that requires the crew to address situationally relevant items critical to that particular takeoff can be inserted during times with lower levels of workload. By having the takeoff brief address important conditions related to the airport, weather, and performance, crews increase their awareness of those conditions affecting takeoff. Having the briefing scheduled for periods of lower workload prior to taxi helps improve situation awareness and decision making during a critical phase of flight.

The successful implementation of ACRM has demonstrated that CRM procedures can be integrated within SOP providing a solid structure to crew management training and assessment. These procedures helped crews form a set of predictable behaviours that increased crew co-ordination and communication. During crew assessment, the procedures helped instructors and evaluators brief and debrief the technical and CRM performance more objectively because procedural performance was more focused than the traditional evaluation of general CRM markers.

Resource management and automation

A generative process, one similar to that used to implement ACRM, can be very effective in producing positive changes in crew performance (Holt *et al.*, 1998), but as can be seen under the subsection on organisational implications, this comes at a cost in personnel and resources and the need for increased co-ordination. Airline can benefit from this cost if they integrate their automation training and procedures using a process such as ACRM. Just as AQP encourages the integration of CRM with technical training, airlines should use efficient means to combine their automation and resource management training and implementation.

Crew training is one of the major costs of the ACRM process. One way to reduce training costs is to combine CRM with elements of technical training, further coupling that with automation training. Part of automation training could be incorporated with existing CRM and technical training, providing real benefits. First, under an integrated curriculum, crews would receive more realistic training that approximates a greater range of operational challenges and conditions. Pilots, whether in the classroom, working with Computer Based Training (CBT), or in simulators, would spend less time memorising and recalling separate automation, CRM, and technical concepts, and would spend more time practising problem solving and performance tasks that develop a mix of management and technical skills. Second, better integration across development and training would reduce the need for independent development efforts. Airlines should facilitate integrated development teams rather than foster the "empire building" that can take place as separate departments compete for funding and personnel. This does not to suggest that there is no need for experts in automation or CRM. As FAA (1996) recommends, airlines should "re-balance" their training investments in favour of automation as it becomes more critical to crew performance. Included in that recommendation should be a warning to airlines to achieve that re-balance in an integrated manner.

There are concrete signs of a preliminary integration of automation with resource management training. Some airlines have recently included automated flight deck skills as part of the curriculum and assessment of CRM skills, but airlines still face many challenges in this area. The increasing amount of flight deck automation has changed the number and nature of operating procedures, but neither the airlines nor the research community understand the effect of these changes on crew performance. Airlines also need to work on combining the assessment of automation with other forms of crew performance. One solution may come from a process such as ACRM, where Line Operational Evaluation (LOE) was used to present crews with problems in order to assess specific aspects of CRM as well as technical performance. Airlines could use this form of evaluation to include more precise forms of automation assessment. In LOE development, airlines can specify the technical, automation, and CRM objectives for the evaluation and include the proper conditions to ensure an integrated and realistic assessment of crew performance. The ACRM process can provide a set of steps toward integration, but an implementation framework is required to guide the primary level of an organisation that need to interact in order to ensure a successful effort.

Dimensions of crew performance

As the role of job context is more recognised in research on job performance (Arvey & Murphy, 1998), automation and resource management integration must address the key levels of that job context. From a performance perspective (Guzzo & Dickson, 1996), those levels are the individual operator, the team, and the organisation. In the airline environment, that translates to the pilot, crew, and organisational levels (see Figure 1 for the three levels).

The next three sections address resource and automation management skills from the perspective of each of these levels of the implementation framework.

The individual pilot level

Traditional automation and parts of resource management training have focused on the individual pilot knowledge and skill requirements. As emphasis has shifted to crew and organisational considerations, current training may overlook the some aspects of the pilot level. The individual pilot will remain important because training task analysis continues to be conducted at the captain and first officer level. Any new automation or resource management skills will require slinks back to those individual task listings. Further, both automation and resource management skills have substantial cognitive elements. With the most established cognitive analysis methods still at the individual level, many of those skill analyses are likely to be performed at the pilot level. Some resource management training takes place at the crew level, but overall, the focus will remain at the pilot level until more effective team analysis methods have been successfully implemented (Seamster & Kaempf, in press).

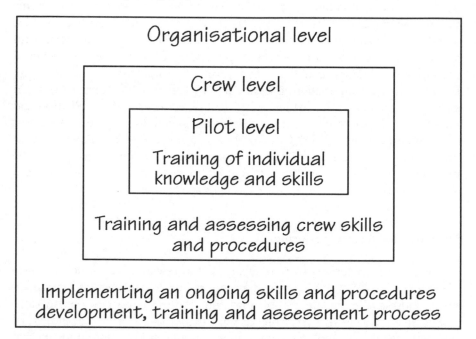

Figure 1: Representation of the levels of job context

A closer look at these reasons emphasises the importance of the individual level in the training of knowledge and skills as well as in the ultimate implementation of an automation and resource management program. Not only has the individual pilot level been the basis for curriculum development, but a substantial part of airline ' training programs still emphasise the pilot. In addition, the pilot level provides a common ground between the traditional task analysis and the more precise cognitive analysis results. Cognitive task analysis has moved from a supplemental tool (Tannenbaum & Yukl, 1992) to a robust set of methods well suited to analyse the more complex task elements found in aviation (Seamster, Redding & Kaempf, 1997). With established cognitive methods emphasising the individual level of analysis, their results best fit at the pilot level. That individual pilot level still plays an important part in the overall structure of the task listing and in the development of new cognitive skill training related to the management of automation and crew resources.

The approach to airline crew formation also tends to emphasise the individual. Within most airlines, crew formation is of limited duration lasting an average of three or four days. Therefore, even when crews are trained together, the training is not aimed at just improving that specific crew, rather it is aimed at helping the individuals to perform in any crew environment (Guzzo & Dickson, 1996). The individual pilot level, although sometimes lost in the rush to consider the crew and organisational level, is now, and will remain an important dimension of automation and resource management skills and will continue to be a substantial area of focus for knowledge and skill training (Seamster & Kaempf, in press). The individual skill level should serve as the starting point for knowledge and skill training to ensure that the resulting skills fit into existing worker task structures, and if needed, that they can be subjected to additional cognitive analysis.

The crew level

Crews are a form of work group with some unique requirements when compared to the individual pilot. One distinguishing characteristic of a crew is that it is structured as a team, and as such has both individual participants and team accountability (Katzenback & Smith, 1993). At airlines, where flight deck performance is highly co-ordinated, accountability lies increasingly with the crew. In the context of automation and resource management, a crew comprises two or more pilots who have complimentary skills, a common set of performance goals and standards, and who act with mutual responsibility and accountability (Katzenback & Smith, 1993). The crew level has become more important with the increasing emphasis on team accountability. In the US pilots were traditionally assessed as individuals, but under AQP, resource management performance assessment focuses on the crew (Holt, Meiman & Seamster, 1996). This growing emphasis on the crew is starting to be seen at some US carriers where automation skills associated with communication, judgement, and monitoring are being assessed.

As performance assessment concentrates more on the crew, the crew management skills need to be more fully addressed both in training and in debriefing and assessment. Without a deeper understanding of crew management skills and how to assess and debrief them, instructors will have limited success improving crew performance. Instructors can usually point out what is wrong with crew performance, but they cannot specify the optimal training intervention to improve that performance (Seamster & Kaempf, in press). For example, an evaluator may tell a crew that they did not attend to a specific task in a timely manner. Such feedback tells the crew that there was a problem, but it does not specify what the crew or individual pilots should practice to better perform under those or related conditions in the future. The crew should be given strategies that will help them start the task earlier or assign tasks to different crew members. Further, the crew should be given the opportunity to practice those strategies within a training or simulator environment. These are elements of diagnostic feedback essential to good skill training. That type of feedback can only be provided based on a deep understanding of crew automation and resource management skills.

In the LOFT and LOE environment, CRM debriefings are generally limited to what the crew did well and less well, and they normally do not specify how the crew can improve performance nor do they provide focused skill practice. This is not the case in manoeuvre validation sessions where specific skill practice is given for technical performance problems such as difficulty with V1 cuts. When a pilot has problems with a specific manoeuvre, the individual is given specific instructions or techniques on how to perfect the manoeuvre. Then the pilot practices the full manoeuvre until he or she reaches the desired level of proficiency. That same level of feedback and practice is not available for automation or resource management problems that occur in training sessions. In the case of resource management difficulties, the instructor and crew may discuss what was done incorrectly, but the crew rarely has the opportunity to practice and refine specific management skills. In the case of automation management problems, when instructors do not have the time to show crews how to program their way out of a set of incorrect entries, they tend to make the corrections so that the crew can get on with the rest of the training. As crew skills are better understood, those skills should not only receive better training but they should also be taken into account when airlines redesign flight deck procedures.

The organisational level

Increasingly greater attention is being paid to the effects of the organisational on job performance and skill development. Organisations have been involved in the identification and specification of skill standards. On the research side there have been studies of the organisational elements affecting job performance (Arvey & Murphy, 1998) looking among other things at the effects of the social and organisational environment on skill acquisition (Voss & Wiley, 1995). Researchers are starting to address the fact that skill acquisition takes place within an organisational environment that can foster or inhibit the need for specific skills. This is particularly true at airlines where operating philosophy and reward systems has a strong influence on training and motivating the development of individual and crew skills.

Within the US, the Goals 2000 legislation established a Standards Board that has identified groupings of industries to define national skill standards (Sheets, 1995) that support worker training, certification, and transition assistance. The skill standards movement has generated new ideas about the nature and role of skills within organisations. To highlight the new perspective, Merritt (1996) compares the traditional *skill-components model* to the newer *professional model*. Under the skill-components model, workers had limited roles with their effectiveness based on how well they performed on a set of tasks. Managers controlled the identification and development of skill standards which were developed by an analyst or Subject Matter Expert (SME). The professional model is very different assuming that the workers are in charge and are able to make decisions. With the professional model, the worker is in control of the work, and tasks are just part of the overall job which includes problem solving and decision making. Organisations need to shift from a top-down to a bottom-up approach if they want to utilise the professional model by involving pilots and crews in developing skill standards and flight deck procedures.

The skill standards legislation has not had a direct impact on airlines, but it highlights positive changes that are currently taking place in other organisations. Some airlines are moving away from using the single SME for their task analyses to multiple SME (Seamster & Kaempf, in press). SMEs are being recruited from key departments and from the required areas of expertise to include automation management, CRM, flight deck procedures, and airline policy and philosophy. The organisation has a strong effect on the overall process for developing and implementing procedures and training. To be most effective, the organisation should become less of a director and more of a facilitator ensuring all levels participate.

The organisation should facilitate the integration process by being responsive to the primary levels within the airline. The organisational level should play a central part in ensuring that the automation and management skills and procedures are properly developed, trained, and assessed. Further, the organisational level facilitates the development of appropriate philosophy, policies, and reward systems to ensure that pilots and crews are properly motivated to develop and maintain high levels of performance.

Integrating resource and automation management

The integration of automation training with resource management and technical training can become a complex and near impossible undertaking unless an airline has an implementation framework to structure the process. Without a framework to ensure that all key elements are included in the implementation process, airlines are likely to adopt one or more products or procedures without establishing an ongoing process essential to successful integration. The framework just specified allows airlines to concentrate the skill training and assessment part of integration at the pilot and crew levels while ensuring that the procedures development and overall implementation is properly embedded at the organisational level. The framework helps to assure that the integration process is facilitated at the organisational level relying on substantial pilot and instructor input in developing effective integration.

There is preliminary evidence that airlines are starting the process by linking crew automation skills with resource management skills. There is also evidence that integrating resource management with operations does improve crew performance (see Holt *et al.*, 1998). Even though these are just pieces of the overall process, they are significant steps that demonstrate how airlines can start developing this ongoing implementation process.

Linking automation and resource management skills

Airline curriculum developers, instructor, and evaluators are placing increasing importance on CRM skills. Technical flight skills and knowledge have formed the basis of crew training, and in the past five to ten years, there has been a growing recognition of the role of CRM skills. These are the skills associated with decision making, situation awareness, and resource management. With growing recognition of the cognitive skills, evidence is surfacing that neither the airlines nor the research community has a consistent definition of (Lanzano, Seamster & Edens, 1997) nor agreed-upon methods (Seamster, Prentiss & Edens, 1997) to identify and analyse these cognitive automation and resource management skills.

Although the research community is unlikely any time soon to agree on the definition of skill, the airlines need a skill definition that will help them in training and assessment. Proctor and Dutta (1995) provide a starting point for such an operational definition, "Skill is a goal-directed, well-organised behaviour that is acquired through practice and performed with economy of effort" (p. 18). This definition provides airlines with two important operational considerations: 1) since skills are acquired through practice, they are likely to be best trained via a practice and feedback cycle, and 2) since they are performed with economy of effort, they can be assessed by observing the quality of performance. This is just a starting point, but it provides airlines with guidance on how to train and assess not only psychomotor skills but the cognitive managerial skills as well.

One major US airline has specified that a management skill has the following properties:

♦ It is assessed through observable behaviours.
♦ It is a measurable level of proficiency to perform a task.
♦ It requires practice to meet a standard of performance.

+ It is directly related to a knowledge of one or more CRM components.
+ It improves individual performance in a crew setting.
+ It enhances mission awareness.

That airline then evaluated a list of 320 potential management skill statements. They developed an instrument for SMEs to specify which skill criteria were met by each of the skill statements. SMEs represented a number of organisational concerns including human factors, fleet aircrew training, as well as individual pilots. SMEs worked individually to determine the degree to which the skill statements met their criteria for management skills. The resulting list of likely management skills was refined by eliminating duplicates and combining lower level statements. Finally, SMEs grouped those statements into the following management skill categories:

+ Communication.
+ Leadership.
+ Situation Awareness.
+ Decision Making.
+ Automation and Technology.

The specific skills statements for the automation category were as follows:

+ Change level of automated system (up or down) to increase situational awareness and avoid work overload.
+ Establish guidelines for PF and PNF duties for the operation of automated systems.
+ Maintain an awareness of the automation modes selected by crew or initiated by FMS.
+ Plan and brief automation modes and configurations.
+ Plan workload and allow sufficient time for programming tasks.
+ Limit programming during critical phases or conditions of flight.
+ Verbalise entries and changes to automated systems.

The above skill statements combined with related Communication, Situation Awareness, and Decision Making skills, have provided that airline with an integrated automation and resource management skill listing.

Implementation considerations

This section presents the major implications of integrating automation and resource management with operations (for a more detailed description of the actual steps, please see Seamster *et al.*, 1998). The section starts with the organisational implications with an emphasis on the considerations for establishing an ongoing process across the organisation. Then crew implications are presented concentrating on management skills training and assessment. Finally, the importance of individual pilot involvement in and input to the skill and procedure development process.

Organisational implications

A major lesson learned from the integration of management skills with SOP was the need for the organisation to establish an ongoing process and not a one-shot training and SOP package. The integration process should provide a set of steps for improving crew performance that is reusable so that additional refinements can be identified and developed with involvement of the entire organisation.

Although the need for integration may originate from the training department, links should be established with other departments to ensure the ongoing viability of the program. The main links that should be established and maintained are between the integration development group, the customer fleets, and the standards department. Whether formal or informal, it is important to have Cupertino from the different departments that will be affected by the integration process, both on the training and procedures side of the organisation.

The preliminary organisational presentations offer one of the first ways to establish links throughout the organisation. Once the link has been established, open communication needs to be developed and maintained. This should be encouraged by the development team, both formally and informally. The formal communication should include scheduled reviews allowing the fleet and standards representatives to see and comment on intermediate versions of any new procedures and training materials. Other formal communications may include scheduled briefings and meetings to ensure that other departments are informed of progress and issues in the development process.

Informal communication links are also important, especially for maintaining flexible links throughout the organisation. During the formal communication process, identify individuals who are interested in the integration process and who understand the advantages of Cupertino between departments. Develop informal links by inviting interested individuals to working meetings where issues are discussed. These informal links should not require additional time or meetings on the part of the development team, and they provide good input and Cupertino from the different departments. In addition, identify a champion within the higher levels of the organisation who can lend support to the program at key periods such as initial funding and during implementation.

Once organisational involvement has been established, steps should be taken to further develop and ensure ongoing management, union, and Principal Operations Inspector (POI) commitment. Each airline has a different management structure and a unique relation with union and POI, but some common steps can be taken to build a strong commitment for the integration process. On the management side, develop an understanding of what will be achieved thought the integrated training, and then provide regular reports to show progress and trends. In developing the preliminary understanding, it can be helpful to provide data on current trends in training, and if automation and resource management data is not available, use technical data such as that from manoeuvres to demonstrate the type of data that will be collected. At some airlines this may be the first program to carefully measure crew performance. In such cases, management should be familiarised with the methods that will be used to collect crew performance data and the types of questions that can be answered through the analysis of that data. Use the commitment-building presentations to identify the types of reports most useful to management. One important outcome of this process is to develop strong commitment for the integration process from top management on down.

On the union side, consider getting one or more union representative involved in the development as early as possible in the process. In the preliminary meetings it should be established that the integration of automation with resource management offers comprehensive training that is well integrated with the technical. Further, crew assessments under this approach are objective, based on SOP, and well specified based on observable crew behaviours. Work closely with at least one representative, and invite them to fly or ride the new LOE or LOFT. Also invite them to be present at the new training courses and at crew assessment sessions such as the LOE or line check. Invite union feedback and keep them informed as the program develops. In the US, the working relationship with the FAA's POI is also very important. Some airlines have developed ambitious new training programs only to find that the POI either does not understand the program or does not see the need for change. The POI should be included, informed, and consulted as soon as is practical within the specific organisation. If possible, make the POI part of the process. Take the time early in the process to explain the need for the new training and make explicit the approach that will be taken.

Another key to establishing an ongoing process is to ensure instructors are informed, practised, and comfortable with the new training. Experience has shown that certain activities help develop instructors into a better training and assessment team. These activities include having training sessions where instructors establish a good level of agreement in their assessment of crew performance. Accurate and timely feedback should be given to instructors prior to and throughout the implementation of an integrated training program. In addition, standardisation meetings should be planned to allow the instructors to voice problems with the process and to work as a team to identify solutions to those problems.

Finally, ongoing successful implementation requires the reporting and use of crew performance data. Once an airline has established that they are collecting reliable and stable crew performance data, they should start reporting trends to appropriate departments within the organisation. Different types of data and formats should be used when reporting to the crew, to the instructors, to fleets, or to management. When properly reported, that data will direct changes or additions to the automation or resource management procedures as well as modifications to the training.

Airlines realises the importance of management training, especially at a general level. That is, they understand that many accidents in commercial aviation have a human factors or crew management component. It is more difficult to understand the relationship between their own airline incidents and automation/resource management. Most airlines do not report such causes of incidents, so there can be a tendency to concentrate on a crew's technical problems. The integration process should bridge this gap in understanding over time to help the organisation understand automation and resource management problems that impact daily operations. The precise methods used to develop this understanding will vary from one airline to another, but there are two important elements. First, the training department should establish and clearly represent to the organisation the link between the new integrated training and a crew's development of better management skills. Second, a clear understanding of the ongoing, feedback-driven nature of the process should be developed. These two elements are essential in establishing a good working environment throughout the organisation.

Crew implications

Crews are a major beneficiary of the integration of automation with resource management in that they gain a standard, proceduralised approach across all flight deck operations. When properly designed and implemented, procedures promote a predictable form of crew co-ordination that is shared and understood by all pilots. This results in a more standard performance that helps crew members participate in planning, decision making, and situation awareness. Research has shown that predictable patterns of interaction, especially in the area of crew communication (Kanki, Lozito, and Foushee, 1989) are associated with better performing flight crews. It has been suggested that when communication is more predictable it tends to be more reliable and more likely to succeed. Integrated management training and procedures promote that standard crew communication and co-ordination which should result in overall improved performance.

Integrated training should produce some noticeable improvements for crews over traditional modular training. Individual implementations may highlight some different improvements, but there are general benefits that should be brought to the crews' attention. A major improvement is the specificity and operational relevance that automation and resource management procedures bring to what may have been a set of general concepts. The new procedures present specific management steps for crews to follow under certain flight conditions. Encourage crews to ask questions at any point through the presentation of the new procedures to ensure that those procedures are clear to all crews.

The integration process provides crews with an operational environment in which they are encouraged to improve their management along side with their technical performance. The process, with its emphasis on continual development, allows the crews to further identify management procedures and training that can improve crew effectiveness.

In addition the integration of automation/resource management with operations has profound implications on the way that crew performance is assessed. A detailed simulator-based assessment should be developed to collect crew performance data after the new crew training has been implemented. This form of crew assessment, based on the LOE, allows for the collection of performance data under a carefully designed and controlled set of flight conditions. Crews should expect a fair and even assessment from all instructor/evaluators, and the evaluators are consistently working to improve the accuracy of crew performance assessment. Crew training should include one or more detailed modules on this new form of assessment, and the crews should be given substantial opportunity to ask questions and to fully understand the improvements in this area.

Finally, assessment under the integration process involves a set of standards, and crews should be presented with the standards along with an explanation of how the standards will lead to a more systematic form of crew assessment. With the team approach to crew performance, all members should work together to improve crew effectiveness. Individual pilots are now expected to work as a team to support each other in working with the new CRM procedures and in developing a set of automation and resource management skills.

Individual pilot implications

The implication for pilots is the need for their understanding and involvement in identification of automation and resource management skills and procedures. Pilots should understand that management skills should be trained in a task-specific context providing crews with practice and feedback. Because automation and resource procedures require crews to perform specific actions under certain conditions, the procedures facilitate skill development when they are trained through practice and feedback. For example, the basic form of a procedure, such as a required statement about automation intentions, can be fairly simple. The skill comes into play when a crew can state those automation intentions quickly and effectively under a variety of conditions and in a way that improves crew effectiveness.

Individual pilots should also be involved in the integration process and can help in the following development steps:

♦ Identifying own airline automation and resource management training needs.
♦ Specifying gaps in existing procedures and documents.
♦ Providing feedback on the prototype new procedures.
♦ Providing SME input to the crew training course development and LOFT/LOE development.

Pilots should be encouraged to help the organisation identify training needs. They should also participate in the specification of gaps and problems with existing procedures. Once preliminary procedures or training has been developed, pilots should be involved in the prototype evaluation. Prototype evaluation is an important part of the refining process, where the development team interacts with a range of users to determine the best form and content for the new procedures or training. This should be iterative, with the feedback from each review being incorporated into the design to achieve one or more procedures that will be adopted by the users to improve performance.

What is required for this step is a paper prototype of the actual checklist, brief guide, QRH, or training material. Even if the final form will be an electronic display, such as the electronic checklists, it is not necessary to develop the prototype in the final media unless it can be done relatively easily and efficiently. The prototype should reflect the content and format of the proposed procedure or training and, in some cases may include alternative representations. Documentation for the prototype should also include an explanation of why it is being proposed and what individual pilot or crew performance problem is being addressed.

Individual pilots can provide informal comments during the presentation, and more formal, quantitative feedback through a form that can be handed out or administered during the user feedback sessions. The user feedback sessions should be designed and scheduled so that the development team can collect a fair amount of data in an efficient manner. Ideally, airlines would be able to perform usability testing, where the procedures are used by pilots in some operational or simulated context. That is not possible in many cases, so working with prototypes in small groups provides a good alternative. If possible, the sessions should be scheduled for five to ten pilots who are likely to work well together. Working with too small a group (less than five) is less efficient, and the individuals are less likely to be stimulated by a wider range of comments. Working with too large a group (substantially more than ten) is more difficult to manage, and the feedback will likely cover a broad range of topics but not in depth.

The feedback sessions should have several parts. The first part can be a more general presentation of the procedures or training material along with their rationale. The presentation can be followed by pilot questions and general comments. Once the group has a good understanding of the material s and their purpose, a form can be administered to ask for comments or ratings about the different aspects of the procedures. Items may include questions about the effects of a proposed procedure on workload, ease of understanding the procedure or materials, how the procedure interfaces with the rest of the SOP, possible problems, and how the material might be improved. In most cases individual feedback, and not consensus, is most useful, so the forms should be completed individually and not as part of a group discussion.

Individual pilot involvement should be facilitated throughout the development and implementation process. As the end users, their comments and feedback is essential to successful integration of automation and resource management skills and procedures.

12 Automation Policy or Philosophy? Management of Automation in the Operational Reality

ÖRJAN GOTEMAN

Scandinavian Airlines, Sweden

Introduction

As aircraft operators, we struggle with the appropriate use of flight deck automation during our line operations. What are good and useable guidelines for the operation of automation on the line? How do we direct the development of procedures and training for computerised flight decks? How do we influence the procurement of new additional equipment, such as Head-Up-Displays for automated flight decks? Most carriers have the desire to streamline and standardise automation procedures, as they have with all their operational procedures. However, this can be difficult because of the lack of standardisation across aircraft cockpit types. The basic recommendation today is therefore that airlines should create, or adopt an automation philosophy that guides their flight deck procedures, training and equipment procurement (FAA, 1996).

Existing philosophies have been criticised for being too vague and of little practical use (ATA, 1998). On the other hand, more specific guidance runs into trouble when it encounters fundamentally different cockpit designs, which typically co-exist within one carrier's fleet. This chapter discusses the lessons we as an operator have learned in how to make an automation philosophy work in our everyday operational and organisational life. In our case, this has led to a practical distinction between an automation philosophy and automation policy. The lessons we have learned are useful for other carriers struggling with similar issues.

The need to live with existing flight deck designs

As discussed in many places in this book, existing flight deck designs are not perfect and suffer from inadequate feedback, keyhole effects and complexity. Nonetheless, these flight decks are certified and are extensively used world-wide, and will be for decades to come. The only economically realistic option for us as airline is to use existing flight deck technology in a wise manner. In other words, our only option is to learn to live with it.

The FAA Human Factors team has identified some measures that could improve the co-operation between flight deck crew and automation on automated aircraft (FAA, 1996). A cornerstone was to formulate and adapt an automation philosophy that can guide the development of procedures, training and equipment procurement. This perspective seems to be well accepted amongst US operators. Since the early 90s, several North American airlines have clearly stated in their operations manuals a philosophy on how automation should be used in their airline. What is remarkable about these operational philosophies is that they specifically give authority to the pilots to choose what they believe is an appropriate level of automation. This is a shift from the not so clearly formulated philosophy of the 70s and 80s, where pilots were expected to always make maximum use of the automation available. The latter expectation, now commonly modified to allow hand-flying in order to practice flying skills, is still more or less the norm in the rest of the world.

It is interesting to note that this shift - towards a statement in the Operations Manual about discretionary use of automation - first occurred in the United States. There may be two reasons for this. The first is rather cynical: if the pilot is allowed to choose what he believes is the right level of automation, he is the one who can get the blame if anything goes wrong. The other reason is one of culture. The prevalent culture of the United States scores high on individuality and low on power distance (Hofstede, 1991). This is associated with a more discretionary use of automation where it is natural to see automation as one of many tools, with which to accomplish an operational goal.

Requirements for an automation philosophy

A weakness with current automation philosophies is that they are aimed at too broad a group of practitioners. This easily makes them too general for practical work that has to be carried out on the flight deck, and gaps and operational problems persist. For example, reports from check airmen indicate concern about a tendency among less experienced younger pilots to solve flight path control problems via the automation, rather than flying manually. Incident reports similarly indicate that pilots still choose levels of automation that are inappropriate given the circumstances (ATA, 1998). Typically they stick with levels that produce a high programming workload. Alternatively, they turn the automation off completely, in cases where an intermediate level would have been beneficial. This points towards a need for further, i.e. more specific guidance among levels of automation.

Standardisation

The aviation industry regards standardisation as canonical and an effective method of enhancing flight safety. Here it implies crossfleet standardisation of procedures, as well as ensuring that all pilots follow the same procedures. The underlying thought is that with less variation in procedures there is less encouragement in pilots variation of practice (Degani & Wiener, 1994). It is also easier to supervise the standard of pilots if all are supposed to react similarly to the same kind of problem.

The problem is that only a few large airlines have a fleet consisting of only one aircraft type. Even the same aircraft type may be equipped differently. The classic approach is to standardise on one type, often the most common aircraft type in the company. But due to flight deck incompatibility, a more common approach is to standardise on generic items while at the same time accepting some aircraft specific procedures - the latter ones being cases where standardisation would create an extra burden on pilots (Degani & Wiener, 1994).

Transfer of pilots

Transfer of pilots to new aircraft types affects the automation policy in two ways. Positive transfer of knowledge obviously eases the transition and gives the pilot an even larger stock of rule-based knowledge adding to the pilot's expertise (Reason 1990). However, negative transfer can occur if the pilot is not trained thoroughly on the new aircraft's technology. Incorrect mental models will increase the risk of reversions to earlier (and now inappropriate) behaviour. Reversion to well-known memory items in out-of-normal, rapidly developing situations from the former aircraft can be hard to eradicate.

Transfer of pilots between aircraft types is obviously a good argument for an automation policy that states that all aircraft in an operator's fleet should be operated in the same manner. The other side of the coin is that there is a cost in lower operational efficiency and increased flight deck workload if the procedures are not consistent with flight deck technology. The extreme is a total standstill of development as all technology and procedures are standardised towards the earlier aircraft types.

Philosophy or policy?

So there is a trade-off. On the one hand, airlines need more specific guidelines for pilots to choose between levels of automated flight. On the other hand, airlines require the more general philosophy that can guide the creation of procedures, the design of training and the procurement of new and additional equipment. To solve this, a differentiation between philosophy and policy is suggested (see also Degani & Wiener, 1994). This chapter reflects this distinction, where both the philosophy and the policy are generic, guiding aircraft type-specific procedures.

Levels of automated flight

Traditionally, automated flight control systems were perceived and described as separate entities. Procedures that evolved from the introduction of new automation were designed accordingly - essentially as single procedures of how to operate each single system. This no longer works because of the intricate coupling between the various automated systems. The addition of new equipment (for example Head-Up Display guidance systems) even exacerbates this problem: are HUDs just a new form of flight director, or are there more subtle interactions between HUD, the other forms of aircraft automation and the flightcrew?

Sheridan (1978) proposed a model of possible levels of allocation of decision making tasks between humans and computers (see also Dekker and Woods in this volume). Although this was almost purely descriptive, the notion of levels has been maintained. Billings (1997) proposed a control-management continuum consisting of six levels for aircraft control automation. The view of automation in levels is echoed in the FAA Human Factors report (1996) as well. This report talks about "what is an appropriate level of automation", rather than what specific subsystem is appropriate in a specific operational context.

Airbus Industrie also uses the term level of automation in their "Golden Rules" (Airbus Flight Training leaflet, no year). Postulating levels of automation certainly makes a useful point: there are different levels, or depths, of human involvement in the control of the process. This makes it possible to translate these levels into teachable procedural guidance. Seeing the automation as a series of levels of human control involvement is also consistent with the way in which pilots have come to view their automated cockpits, and flows into a straightforward policy: be proficient at all levels.

Definition and terminology of levels

But parsing automation up in certain levels only goes a small way towards an operational philosophy and policy. From the operator's standpoint, levels of automation are oversimplified as well as unspecific. Various intertwined problems make it difficult to translate them into practical guidance for pilots in automated cockpits. First, postulating levels of automation ignores interlevel couplings - an increasingly prevalent feature of complex flight deck systems. Second, it is fine to state what a computer might or might not do at any particular level, but what about the human? Specifically, if you have two operators interacting with the system at the same time, what are their respective tasks and roles going to be at the various levels? How do we preserve the traditional system of pilot cross-checking across automation levels?

Sheridan and Billings make no suggestions. But thinking through this and making it explicit has been critical for our automation policy to achieve pilot acceptance. The basic conclusion is that at different levels there are different things to do for each human crewmember, and specifically different things to do to support and double check the other crewmember. One complication here is that it is not uncommon to have different levels of automation simultaneously responsible for the two axes (vertical, lateral) of flight path control. A typical example is a non-precision approach, where the highest level of automation (LNAV) governs the lateral or horizontal path, but a lower level of automation (V/S) controls the vertical movement of the aircraft. The question is, which of these levels determines the division of human roles and work in the cockpit - the higher or the lower level? After much deliberation we selected the lower level, since that generally puts the pilot-not-flying (PNF) in a more supportive role relative to the pilot-flying (PF).

Most airlines that have postulated levels of automation have some criterion for when to switch crewmember roles, most notably when the need arises for lookout or support duties. We have decided to make the automation level itself the criterion on which crewmember roles are based. As soon as a level of automation is reached where the PF controls the aircraft's flight path without having to hand-fly, (s)he is also responsible for autopilot inputs. This includes flight management system inputs. When the PF is handflying, the PNF is responsible for autopilot manipulations (including the FMS), but those are discouraged at lower altitudes or higher workload phases of flight, since the PF is then hardly in a position to double-check the computer entries that the PNF makes on his CDU. A question in that case would also arise: if the PNF makes flight-path determining computer entries, is the PF really still PF - in other words is the PF technically the "flying" pilot? Obviously, the demarcation between PF and PNF is blurred and perhaps the labels have lost their clarity relative to earlier cockpit designs. The loss of labels is not necessarily to be lamented. But flight safety is dependent on flight deck discipline. And this relies in turn on clear procedures for who does what when - regardless of what the label will be (for example: pilot managing and pilot not-managing?).

Transitioning between levels

Transitioning between levels is not always straightforward either. To disconnect the autothrust on an Airbus A320, a pilot has to first align the non-moving throttles with what thrust the engines are producing at that moment in order to prevent unwanted thrust changes during disconnect. Training pilots to transition smoothly and efficiently between the levels, while still having the whole crew in the loop, is critical if airlines are to reduce the risk of automation surprises and bumpy transfers of control.

Scope of an automation philosophy

An automation philosophy is not only linked to existing equipment. It should also be useful for an airline's overall procedure construction, training development and equipment procurement and it should not boil down to a set of detailed procedures. These procedures may have to change with the arrival of new flight deck technology while the philosophy remains the same. It must also be consistent with the cultural context, in which the airline operates. Again, it is not a coincidence that the North American carriers were the first to accept an automation philosophy giving the pilots authority to use automation at their discretion. (Sherman, Helmreich and Merrit, 1997)

Manual tree

In order for an automation philosophy to be stable, it must be written down. In order for an automation philosophy to be used, it must be communicated. The aviation industry is very strongly regulated and the operational procedures must be documented in an operations manual. How the operations manual is constructed will affect the description of the automation philosophy.

The Joint Aviation Authorities (JAA) in Europe have described how a preferred operations manual of an operator should be organised to facilitate the approval from the authority. (JAR-OPS 1, 1997) Here the JAA suggests a division in the operations manual in Part A, General and Part B, Type-specific. This division is quite useful and can successfully be applied to an automation philosophy. Indeed it gives a hint how to solve the specific-generic dilemma. A philosophy, together with guidelines in a policy can be described in the operations manual part A, guiding the type-specific procedures in part B. This will not lead to identical procedures across different aircraft types, due to different levels of technology. However, it will lead to procedures which are consistent with the same automation philosophy and policy. It will also facilitate the description of compulsory use of certain equipment during some operations, where this is the only certified means of compliance, such as the use of autoland for CATIII approaches on non-HUD aircraft.

Automation policies need to be expressed in a defined terminology in order to be of any use for standardisation purposes. Deviation from the policy must be measurable during check rides by line check pilots. The concept of automation levels must be translatable into procedures. For example, what are the duties of the Pilot Not Flying when using automation at this or that level? If this is not clearly stated, the policy will not be useful.

Implications of the AP

We must assess the operational and organisational implications of an automation policy before we attempt to implement it. For instance, if airlines blindly follow the recommendations from the FAA and ATA (that pilots should have the authority to choose appropriate levels of automation for the task at hand) there may be objections, especially from airlines that operate within a strongly hierarchical culture. One overall recommendation with respect to the creation and implementation of an automation policy is that fleet captains must trust their own pilots, take into account their cultural and operational heritage, and educate them rather than control them through strict rules in the name of cockpit discipline.

Appendix - SAS Automation Policy

Reprinted with permission from Scandinavian Airlines Flight Operations Manual, FOM:

Excerpts from FOM 3.3.2 –

The use of Checklists and standard terminology

5.   Standard terminology

In all normal and emergency procedures the English language shall be used. The terminology of the Normal and Emergency Check Lists as well as of the Expanded Check Lists shall be adhered to.

The terms mentioned below have, if not otherwise stated, systemwide applicability.

The PF shall during all stages of operation clearly state his planning intentions and actions as the flight progresses. This includes calling out verified mode selections, FMS entries and annunciations which alter the level of automation of AFS.

5.1.   Automatic Flight System

PF shall call out the new mode annunciation as depicted on the FMA when transitioning between levels of automation. E.g. "heading select", "LNAV", "autopilot command A". Any pilot shall callout the mode displayed on the FMA when a discrepancy from anticipated annunciation or aircraft behavior is detected.

FOM 3.2.4. Management of Auto Flight Systems

1.1   Automation Philosophy

Automation is a tool, provided to enhance safety, reduce pilot workload and improve operational capabilities.

Pilots shall be proficient in operating their aircraft in all levels of automation, as well as transitioning between different levels of automation. The pilot shall use what he believes is the most appropriate level of automation for the task at hand with regard to safety, passenger comfort, regularity and economy.

SAS policy on the use of automation for control and guidance of aircraft flight path and speed is described in the subchapters below. It shall guide the development of procedures, training and equipment procurement.

1.2   Definitions

Levels of Automation

The level of automation is determined by how much authority is given to AFS for control the aircraft's flight path or speed. It ranges from minimum possible

AFS authority in basic manual level to maximum possible AFS authority in managed automatic level.

- Basic Manual level.
  The aircraft is hand-flown, usually without Flight Director guidance.
- Guided Manual level.
  The aircraft is hand-flown, following Flight Director or HUD guidance.
- Directed Automatic level.
  The aircraft is flown with the autopilot engaged in modes associated with Mode Control Panel, or Flight Guidance Panel inputs (e.g. vertical speed, heading select, VOR/LOC).
- Managed Automatic level.
  The aircraft is flown with the autopilot engaged in modes coupled to the FMS/RNAV (e.g. VNAV, nav track).

1.3   The use of automation

The level of automation used at any time shall be the most appropriate for the task at hand with regard to safety, passenger comfort, regularity and economy within the limits of respective AFM/AOM. Both pilots shall be aware of intended level of automation. To the extent suitable and as prescribed in AFM/AOM, basic data for the navigation systems shall be used for monitoring of AFS performance. In aircraft with functioning dual Flight Director systems, both FD computers shall be used.

1.4.   Guidance for the use of automation

Basic Manual level is used where immediate, decisive and correct control of the aircraft flight path is required. This includes avoidance /escape /recovery maneuvers. These are essentially non-normal maneuvers and with the exception of intentional basic manual flying this should be considered a transitory level of automation.

Guided Manual level is the normal level when hand-flying the aircraft. The guided manual level is appropriate in low density traffic areas. Autothrottle is normally used.

Directed Automatic level is used where short term objectives are being met. The directed automatic level is normally used in terminal areas and is also a normal transitory level when flying below 10000 feet and pilot workload does not permit reprogramming FMS. Autothrottle is normally used.

Managed Automatic level is the recommended level of automation to achieve long-term objectives. The managed level is normally used in climb, cruise and descent, using FMS programming accomplished pre-flight. The managed level may also be used for departure or approach, provided this procedure is described in respective AFM/AOM and workload permits FMS/RNAV programming. Autothrottle is normally used.

If any uncertainty exists regarding AFS behavior, PF should revert to a lower level of automation.

1.5 Crew coordination

The lowest level of automation used at any time determines allocation of crew duties with regard to AFS.

During engagement/disengagement of autopilot or switching of autopilot, PF shall always have one hand on the control column. During departures PF shall have his hands on the controls. Thrust levers shall be guarded below 1500 AGL. During approach shall PF have his hands on the controls and thrust levers below 1500 feet RH, except for necessary inputs to AFS. The pilot making altitude entries in AFS shall point a finger at the altitude readout until the other pilot has confirmed and acknowledged the setting as depicted in AOM/AFM.

- Programming of AFS on ground is normally the duty of PF.
- At the Guided Manual level, PNF will make the required AFS entries and mode selections upon order from PF.
- At the Directed Automatic level, PF will make the required AFS entries and mode selections.
- At the Managed Automatic level, PF manages the aircraft's flightpath through the FMS and normally makes the required FMS entries and mode selections. FMS entries below 10000 feet other than short commands (e.g. "direct to" entries or speed interventions) should be accomplished by PNF upon order from PF. PF navigational display should be used in a mode which shows the active route and at least the first active waypoint.

Automation level	A/P	F/D & HUD	PF	PNF
Basic Manual	OFF	OFF or not followed	Handles the flight controls	Monitors flight progress Calls out impending flight envelope deviations
Guided Manual	OFF	ON	Handles the flight controls	Monitors flight progress Sets up AFS on PF order
Directed Automatic	ON	ON	Makes MCP/FGP selections Monitors flight progress	Monitors flight progress
Managed Automatic	ON	ON	Makes input to FMS Monitors flight progress	Monitors flight progress

Bibliography

Aeronautica Civil (1996). *Controlled flight into terrain, American Airlines 965, Boeing 757-223, N651AA near Cali, Colombia, December 20, 1995* (Aircraft Accident Report). Republica de Colombia: Aeronautica Civil.

Air Transport Association (1997). *Towards a model training program for FMS-generation aircraft.* Report of the Subcommittee on Automation Human Factors. Washington, D.C.: ATA.

Air Transport Association (1998). *Potential knowledge or policy gaps regarding operation of FMS-generation aircraft.* Report of the Subcommittee on Automation Human Factors. Washington, D.C.: ATA.

Airbus Industrie (No Year). *Golden Rules.* Information leaflet from Airbus Industrie Flight Training, Toulouse.

Amalberti, R. (1997). Automation in aviation: A human factors perspective. In D. Garland, J. Wise & V. D. Hopkin (Eds.), *Aviation human factors.* Hillsdale, N.J.: Erlbaum.

Arbib, M. A. (1964). *Brains, machines and mathematics.* New York: McGraw-Hill.

Arvey, R. D. & Murphy, K. R. (1998). Performance evaluation in work settings. *Annual Review of Psychology, 49,* 141-168.

Baiada, R. M. (1995). ATC biggest drag on airline productivity. *Aviation Week and Space Technology, 31,* 51-53.

Benoist, Y. (1998). *The impact of automation on accident risk.* Airbus Industrie report.

Bent, J. (1997). Training For New Technology - Practical Perspectives. *Proceedings of the 9th International Symposium on Aviation Psychology, Colombus Ohio.*

Billings, C. E. (1991). *Human-centered aircraft automation: A concept and guidelines (NASA Technical Memorandum 103885).* Moffett Field, CA: NASA-Ames Research Center.

Billings, C. E. (1997), *Aviation Automation, The search for a human centered approach,* Hillsdale, N.J: Erlbaum.

Billings, C. E. & Cheaney, E. S. (1981, September). Information transfer problems in the aviation system. In: *The information transfer problem: Summary and Comments,* pp. 89-90. Washington, D.C.: NASA Technical Paper 1875.

Bittel, L. R. (1964). *Management by exception: Systematising and simplifying the managerial job.* New York, NY: McGraw Hill.

. & Lincoln J. E. (1988). *Engineering data compendium: Human perception rformance*. AAMRL, Wright-Patterson AFB.

B. (1988). Organisation for decision making in complex systems. In L. P. Goodstein, H. B. Andersen & S. E. Olsen (Eds.), *Tasks, Errors, and Mental Models*. London: Taylor & Francis.

Brehmer, B. (1991). Modern information technology, time scales and distributed decision making. In J. Rasmussen, B. Brehmer & J. Leplat (Eds.), *Distributed Decision Making: Cognitive Models for Co-operative Work*. Chidester: Wiley.

Broadbent, D. E. (1991). *Human reliability assessment: A critical overview*. Advisory Committee on the Safety of Nuclear Installations Study Group on Human Factors.

Byrnes, R E. & Black, R. (1993). Developing and implementing CRM programs: the Delta experience. In E. L. Wiener, B. G. Kanki & R. L. Helmreich (Eds.), *Cockpit Resource Management*. San Diego: Academic Press, Inc.

Cannon-Bowers, J. A., Salas, E. & Converse, S. E. (1994). Shared mental models in expert team decision making. In N.J. Castellan, Jr. (Ed.) *Current issues in individual and group decision making*. Hillsdale, NJ: Erlbaum.

Chapanis, A. (1970). Human factors in systems engineering. In K. B. De Greene (Ed.), *Systems psychology*, pp. 51-78. New York: McGraw-Hill.

Collins (1997), FMS-4200 Flight Management System, Pilot's Guide.

Cook, R. I. & Woods, D. D. (1996). Adapting to new technology in the operating room. *Human Factors*, in press.

Cook, R. I., Woods, D. D. & McDonald, J. S. (1991). Human Performance in Anesthesia: A Corpus of Cases. Cognitive Systems Engineering Laboratory Report, prepared for Anesthesia Patient Safety Foundation, April 1991.

Cooper, D. E. (1996). *Heidegger*. London, UK: Claridge Press.

Cooper, G. (1994). Euro flow control. *Royal Aeronautical Society Aerospace*, 21 (3), 8-11.

Cotton, W. B. (1995). Free flight in domestic ATM. *Journal of Air Traffic Control*, *3*, 10-17.

Courteney, H. Y. (1996) Practising what we preach. *Proceedings of the First International Conference on Engineering Psychology and Cognitive Ergonomics*, Stratford-upon-Avon, UK.

Courteney, H. Y. (1998a). User Error: Practical Ways to Protect Your System. *Proceedings of the International Council of Systems Engineers Conference on People, Teams and Systems*, Vancouver, BC, Canada, July.

Courteney, H. Y. (1998b). Assessing error tolerance in flight management systems. *Proceedings of the Second International Conference on Engineering Psychology and Cognitive Ergonomics*, Oxford, UK.

Courteney, H. Y. & Earthy, J. V. (1997). Assessing system safety to 10-JOKER: Accounting for the human operator in certification and approval, *Proceedings of the Conference of the International Council of Systems Engineers (INCOSE) at the European Space Agency (ESA)*.

Coyne, J. K. (1995, September 27). An agency that won't fly. *Wall Street Journal*, 11.

Degani, A. S. & Wiener, E. L. (1991). Philosophy, policies, and procedures. *Proceedings of the Sixth International Symposium on Aviation Psychology* (pp. 184-191). Columbus: The Ohio State University.

Degani, A. S. & Wiener E. L. (1994) *On the design of flight-deck procedures* (NASA Contractor Report 177642). Moffett Field, CA: NASA-Ames research center.

Dekker, S. W. A. (1996). Cognitive complexity in management by exception: Deriving early human factors requirements for an envisioned air traffic managment world. In D. Harris (Ed.), *Engineering Psychology and Cognitive Ergonomics, Volume I: Transportation systems*. Aldershot, England: Ashgate, pp. 201-210.

Dekker, S. W. A. & Woods, D. D. (in press). To intervene or not to intervene: The dilemma of management by exception. *Journal of Cognition, Technology and Work*.

Dekker, S. W. A. & Wright, P. C. (1997). Function allocation: A question of task transformation. *Proceedings of ALLFN '97 — Revisiting the allocation of functions issue conference*, Galway Ireland.

Dornheim, M. A. (1995, July 31). Equipment will not prevent free flight. *Aviation Week and Space Technology, 31*, 46.

ECAC (1980). *Document 17*, European Civil Aviation Conference.

ECOTTRIS (1998). *European Collaboration on Transition Training Research for Improved Safety: Final Report*. European Community DGVII (Transport) Contract AI-96-SC.201.

Edwards, E. & Lees, F. P. (1974). *The human operator in process control*. London: Taylor & Francis.

Ehn, P. & Löwgren, J. (1997). Design for quality-in-use: Human-computer interaction meets information systems development. In M. Helander, T. K. Landauer, & P. Prabhu, (Eds.), *Handbook of Human-Computer Interaction*, 2nd ed., pp. 299-313. Oxford: Elsevier Science.

Endsley, M. R. & Kiris, E. O. (1995). The out-of-the-loop performance problem and level of control in automation. *Human Factors, 37*(2), 381-394.

Endsley, M. R. (1988). Design and evaluation for situation awareness enhancement. *Proceedings of the Human Factors Society 32nd Annual Meeting.* Human Factors Society, Santa Monica, CA.

Ephrath, A. R. & Young, L. R. (1981). Monitoring vs. man-in-the-loop detection of aircraft control failures. In J. Rasmussen & W. B. Rouse (Eds.). *Human detection and diagnosis of system failures* (pp. 143-154). New York: Plenum.

Ericsson, K. A. & Simon, H. A. (1980). Verbal reports as data. *Psychological Review,* 87, 215-251.

Ericsson, K. A. & Simon, H. A. (1993). *Protocol analysis: verbal reports as data.* Cambridge MA: MIT Press.

European Coal and Steel Community (1976). Human factors evaluation at Hoogovens No. 2 hot strip mill. Proceedings of the meeting on 26-27 October, Secretariat of Community Ergonomics Action, Europen Coal and Steel Community, Luxembourg.

Ewell, C. D. & Chidester, T. R. (1994). Human factors consequences of aircraft automation. *Flight Deck.* Dallas, TX: American Airlines.

FAA (1989) *Advisory Circular 25-15, Approval of Flight Management Systems in Transport Category Airplanes.* Washington, DC: Federal Aviation Administration.

FAA (1991). *Advisory Circular 120-54: Advanced Qualification Program.* Washington, DC: Federal Aviation Administration.

FAA (1996). *The Interface Between Flightcrews and Modern Flight Deck Systems.* Washington, DC: Federal Aviation Administration.

FAA (1997) *Advisory Circular 120-28D, Criteria for approval of CAT III weather minima Takeoff, Landing and Rollout,* Draft 13. Washington, DC: Federal Aviation Administration.

FAA (1998) *Code of Federal Regulations,* Aeronautics and Space, Parts 1-59. Washington, D.C.: Federal Aviation Administration.

Feltovich, P. J., Spiro, R.J. & Coulson, R. L. (1993). Learning, teaching, and testing for complex conceptual understanding. In Fredriksen, N., Mislevy, R. J., Bejar, I. I. (Eds.), *Test Theory for a New Generation of Tests,* Academic Press.

Fitts, P. M. (Ed.), (1951). *Human engineering for an effective air navigation and traffic-control system.* Ohio State University Research Foundation, Columbus, Ohio.

Foushee, H. C. & Helmreich, R. L. (1988). Group interaction and flight crew performance. In E. L. Wiener & D. C. Nagel (Eds.), *Human factors in aviation.* San Diego, CA: Academic Press.

GAO (1997). *FAA's guidance and oversight of pilot crew resource management training can be improved.* United States General Accounting Office, GAO/RCED-98-7 Human Factors.

Gibbs, W. W. (1995, December). Free-for-all flights: The FAA plans a revolution in air traffic control. *Scientific American, 12,* 11-12.

Good, M., Spine, T., Whiteside, J. & George, P. (1986). User-derived impact analysis as a tool for usability engineering. In *Human Factors in Computing Systems (CHI '86 Proc.),* pp. 241-246. New York: ACM Press.

Gregorich, S. E., Helmreich, R. L. & Wilhelm, J. A. (1990). The structure of cockpit management attitudes. *Journal of Applied Psychology,* 75, 682-690.

Guzzo, R. A. & Dickson, M. W. (1996). Teams in organisations: Recent research on performance and effectiveness. *Annual Review of Psychology,* 47, 307-338.

Hancock, P. A. (1992). On the future of hybrid human-machine systems. In J. A. Wise, V. D. Hopkin & P. Stager (Eds.). *Verification and validation of complex systems* (pp. 61-85). NATO-ASI series F: Computer and Systems Sciences, V. 10.

Harris, D. (Ed.) (1997). Human factors for flight deck certification. *Proceedings of the European Workshop to Develop Human Factors Guidelines for Flight Deck Certification,* London, UK. Bedford: Cranfield University Press.

Hart, S. G. & Staveland, L. E. (1988). Development of a multi-dimensional workload rating scale: Results of empirical and theoretical research. In P. A. Hancock & N. Meshkati (Eds.), *Human Mental Workload.* Amsterdam: Elsevier.

Helmreich, R. L., Merritt, A. C. & Wilhelm, J. A. (1999). The evolution of crew resource management training in commercial aviation. *The International Journal of Aviation Psychology,* 9, 19-32.

Hofstede, G. (1991). *Cultures and organisations.* New York: McGraw-Hill.

Hollnagel, E. (1990). The design of error tolerant interfaces. *First International Symposium on Ground Data Systems For Spacecraft Control, Darmstadt,* FRG, June 26-29. (ESA Special Publication SP-308).

Hollnagel, E. (1992). Coping, coupling and control: The modelling of muddling through. *Paper presented at the 2nd Interdisciplinary Workshop on Mental Models,* Robinson College, Cambridge, UK.

Hollnagel, E. (1993). Models of cognition: Procedural prototypes and contextual control. *Le Travail humain, 56(1),* 27-51.

Hollnagel, E. (1997). Designing for complexity. In G. Salvendy, M. J. Smith & R. J. Koubek (Eds.), *Design of computing systems: Social and ergonomic considerations,* pp. 217-220 (Proceedings of the Seventh International Conference on Human-Computer Interaction, San Francisco, CA, August 24-29, 1997). Amsterdam: Elsevier.

Hollnagel, E. (1998). Context, cognition, and control. In Y. Waern (Ed.). *Co-operation in process management – Cognition and information technology*. London: Taylor & Francis.

Hollnagel, E. & Cacciabue P. C. (1991), *Cognitive modelling in system simulation*. Proceedings of Third European Conference on Cognitive Science Approaches to Process Control, Cardiff, September 2-6, 1991 .

Hollnagel, E. & Niwa, Y. (1996) A cognitive systems engineering approach to computerised procedure presentation. *Proceedings of Cognitive Systems Engineering in Process Control (CSEPC 96)*, Kyoto, Japan.

Hollnagel, E. & Woods, D. D. (1983). Cognitive systems engineering: New wine in new bottles. *International Journal of Man-Machine Studies, 18*, 583-600.

Hollnagel, E., Pederson, O. M. & Rasmussen, J. (1981). *Notes on human performance analysis* (Tech. Rep. Riso-M-2285). Denmark: Riso National Laboratory.

Holt, R. W., Meiman, E. & Seamster, T. L. (1996). Evaluation of aircraft pilot team performance. *Proceedings of the Human Factors and Ergonomics Society 40th Annual Meeting*. Santa Monica, CA: Human Factors and Ergonomics Society.

Holt, R. W., Boehm-Davis, D. A., Ikomi, P. A., Hansberger, J. T., Beaubien, J. M., Incalcaterra, K. A., Seamster, T. L., Hamman, W. & Schultz, K. (1998). *CRM Procedures and Crew Performance*. Washington DC: Federal Aviation Administration, Office of the Chief Scientific and Technical Advisor for Human Factors.

Honeywell (1997), *FMZ Series Flight Management System*, Pilots Operating Manual.

Hughes, D. (1992, March 23). Automated cockpits: Keeping pilots in the loop. *Aviation Week and Space Technology, 12*, 48-70.

Hunt, G. F. J. (1991). *Person communication*.

IATA (1994). *Aircraft Automation Part 1: Operation and Safety Considerations and Part 2: Engineering and Maintenance Considerations*. Directorate Flight Operations and Directorate Engineering & Environment, IATA.

ISO 9241 (1997-1998), parts -11, -14, -15, -17, International Standard *Ergonomic Requirements for office work with visual display terminals (VDTs)*.

JAA (1997) *Joint Aviation Requirements*, JAR-OPS 1.

Kanki, B. G., Lozito, S. & Foushee, H. C. (1989). Communication indices of crew co-ordination. *Aviation, Space, and Environmental Medicine*, 60:1, 56-60.

Kantowitz, B. H. & Sorkin, R. D. (1983). *Human factors: Understanding people-system relationships*. New York: Wiley.

Katzenback, J. R. & Smith D. K. (1993). The discipline of teams. *Harvard Business Review*, 71, March-April, 111-120.

Kerstholt, J. H., Passenier, P. O., Houttuin, K. & Schuffel, H. (1996). The effect of a priori probability and complexity on decision making in a supervisory control task. *Human Factors, 38(1)*, 65-78.

Kessel, C. J. & Wickens, C. D. (1982). The transfer of failure-detection skills between monitoring and controlling dynamic systems. *Human Factors, 24*(1), 49-60.

Klein, G. A. (1993). A Recognition-Primed Decision (RPD) model of rapid decision making. In G. Klein, J. Orasanu, R. Calderwood, & C. E. Zsambok, (Eds.), *Decision Making in Action: Models and Methods*, pp. 138-147. Norwood, NJ: Ablex.

Lanzano, J., Seamster, T. L. & Edens, E. S. (1997). The importance of CRM skills in an AQP. *Ninth International Symposium on Aviation Psychology.* Columbus, OH: The Ohio State University.

Layton, C., Smith, P. J. & McCoy, C. E. (1994). Design of a co-operative problem-solving system for en-route flight planning: An empirical evaluation. *Human Factors, 36(1)*, 94-119.

Lee, J. & Moray, N. (1992). Trust, control strategies and allocation of function in human-machine systems. *Ergonomics, 35*(10), 1243-1270.

Lee, W. (1971). *Decision theory and human behavior.* New York: John Wiley & Sons.

Lehman, C. (1998). We need a new jet transport pilot training model. *Journal for Civil Aviation Training, (8)*, 7, 34-39.

Lerner, A. Y. (1975). *Fundamentals of cybernetics.* London: Plenum.

Leslie, J. (1996). Free flight: The solution to the antiquated air traffic control system? Make pilots their own air traffic controllers! *Wired, 3*, 92-97.

Löwgren, J. & Stolterman, E. (1998). *Design av informationsteknik: Materialet utan egenskaper* [Designing information technology]. Lund, Sweden: Studentlitteratur.

Mancuso V. (1995). Creating automation excellence. *Presentation at Flight safety Foundation's 48th Annual Safety Seminar*, Seattle.

McClellan, J. M. (1998). Is the sky overcrowded? *Flying (7)*, 13-16.

McGregor, D. (1960). *The human side of enterprise.* New York: McGraw-Hill.

McLucus, J. L., Drinkwater, F. J. & Leaf, H. W. (1981). Report of the President's Task Force on Aircraft Crew Complement. Washington, DC.

Merritt, D. (1996). *A conceptual framework for industry-based skill standards.* Berkeley, CA: National Centre for Research in Vocational Education, University of California, Berkeley.

Miller, G. A., Galanter, E. & Pribram, K. H. (1960). *Plans and the structure of behavior.* New York: Holt, Rinehart & Winston.

Molich R & Nielsen J. (1990), *Improving a human-computer dialog,* Computing Practices, volume 33 no. 3.

Moll van Charante, E., Cook, R. I., Woods, D. D., Yue, L. & Howie, M. B. (1993). Human-computer interaction in context: Physician interaction with automated intravenous controllers in the heart room. In H.G. Stassen, (Ed.). *Analysis, Design and Evaluation of Man-Machine Systems,* Pergamon Press, pp. 263-274.

Moray, N. (1997). Human factors in process control. In G. Salvendy (Ed.), *Handbook of human factors and ergonomics,* pp. 1944-1971. New York: Wiley Interscience.

Moray, N., Lee, J. & Hiskes, D. (1994, April). Why do people intervene in the control of automated systems? *Proceedings of the First Automation Technology and Human Performance conference,* Washington, DC.

Nash, T. (1997). A new approach to ab initio training. *Journal for Civil Aviation Training, 8 (7),* 34-39.

Nash, T. (1998). British Airways pilot recruitment drive. *Journal of Civil Aviation Training, 9 (1),* 33-35.

National Research Council (1997). *Taking flight: Education and training for aviation careers.* Washington, D.C.: National Academy Press.

National Research Council (1998). *The future of air traffic control: Human operators and automation.* Washington, DC: National Academy Press.

National Transportation Safety Board (1986). *China Airlines B-747-SP 300 NM northwest of San Francisco, CA, 2/19/85 (NTSB AAR-86/03).* Washington, DC: NTSB.

National Transportation Safety Board (1994). *Safety Recommendations A-94-164 through 166 (concerning China Airlines Airbus A-300-600R accident at Nagoya, Japan, April 26, 1994).* Washington, DC: NTSB.

National Transportation Safety Board (1997). *Grounding of the Panamanian passenger ship* Royal Majesty *on Rose and Crown shoal near Nantucket, MA, June 10, 1995* (Marine accident report NTSB/MAR-97/01). Washington, DC: NTSB.

Neisser, U. (1976). *Cognition and reality: Principles and implications of cognitive psychology.* San Francisco, CA: Freeman.

New Scientist (1996), *Out of their hands,* 23 November 1996.

Newell, A. (1990). *Unified theory of cognition.* Cambridge, MA.: Harvard University Press.

Newell, A. & Simon, H. A. (1972). *Human problem solving.* Englewood Cliffs, NJ.: Prentice-Hall.

Nolan, M. S. (1994). *Fundamentals of air traffic control.* Belmont, CA: Wadsworth Publishing Company.

Nordwall, B. D. (1995, July 31). Free flight: ATC model for the next 50 years. *Aviation Week and Space Technology, 31,* 38-41.

Norman, D. A. (1989). *The problem of automation: Inappropriate feedback and interaction not overautomation* (ICS Report 8904). La Jolla, CA: Institute for Cognitive Science, University of California-San Diego.

Obradovich J. H. & Woods, D. D. (1996). Users as designers: How people cope with poor HCI design in computer-based medical devices. *Human Factors*, in press.

Orasanu, J. (1994). Shared problem models and flight crew performance. In N. Johnston, N. McDonald & R. Fuller (eds.), *Aviation psychology in practice*. Aldershot, UK: Ashgate.

Orasanu, J. (1995). Evaluating team situation awareness through communication. *Proceedings of the International Conference on Experimental Analysis and measurement of situation awareness*. Daytona beach, FL.

Orasanu, J. & Connolly, T. (1993). The reinvention of decision making. In G. Klein, J. Orasanu, R. Calderwood & C. E. Zsambok (Eds.) *Decision making in action: Models and methods*. pp. 3-20. Norwood, NJ: Ablex.

Parasuraman, R. & Riley, V. (1997). Humans and automation: Use, misuse, disuse, abuse. *Human Factors*, 39, 230-253.

Parasuraman, R. (1987) Human-computer monitoring. *Human Factors, 29*(6) 695-706.

Pawlak, W. S., Brinton, C. R., Crouch, K. & Lancaster, K. M. (1996). A framework for the evaluation of air traffic control complexity. *Paper presented to the American Institute of Aeronautics and Astronautics*.

Perrow, C. (1984). *Normal Accidents*. New York: Basic Books.

Proctor, R. W. & Dutta, A. (1995). *Skill acquisition and human performance*. Thousand Oaks, CA: Sage Publications.

Rasmussen J, (1998) Ecological Interface Design for Reliable Human-Machine Systems, *Journal of Human Factors in Aviation*, in press.

Reason, J. T. (1988). Cognitive aids in process environments: prostheses or tools? In E. Hollnagel, G. Mancini & D. D. Woods (Eds.), *Cognitive engineering in complex dynamic worlds*. London: Academic Press.

Reason, J. T. (1990). *Human Error*. Cambridge, UK: Cambridge University Press.

Riley, V. (1994). A theory of operator reliance on automation. In M. Mouloua & R. Parasuraman (Eds.). Human performance in automated systems: Current research and trends (pp. 8-14). Hillsdale, NJ: Erlbaum.

RMB Associates. (1996). *Free flight: Reinventing air traffic control, volume 2: The "minimalist" solution* (Study and research report jointly produced by RMB Associates and Aviation Systems Research Corporation). Evergreen, CO: R. M. Baiada.

Robertson, M. M. & Endsley, M. R. (1994). The role of crew resource management (CRM) in achieving team situation awareness in aviation settings. *Proceedings of the 21st Conference of the Western European Association for Aviation Psychology*. Dublin, Ireland.

Robertson, S. P. & Zachary, W. W. (1990). Conclusion: Outlines of a field of co-operative systems. In S. P. Robertson, W. W. Zachary & J. B. Black (Eds.). *Cognition, computing and cooperation*. Norwood, NJ: Ablex.

Rochlin, G. (1986). "High-reliability" organizations and technical change: Some ethical problems and dilemmas. *IEEE Technology and Society Magazine*, September 1986.

Rochlin, G. I., La Porte, T. & Roberts, K. H. (1987). The self-designing high reliability organization, aircraft carrier flight operations at sea, *Naval War College Review*, Autumn, 76-90.

RTCA (1995). *Report of the RTCA board of directors' select committee on free flight*. Washington, DC: Radio Technical Commission on Aeronautics.

Ruffell Smith, H. P. (1979). *A simulator study of the interaction of pilot workload with errors, vigilance, and decisions* (NASA TM-78482). Moffett Field, CA: NASA-Ames Research Center.

Sanderson, P. M. (1989). The human planning and scheduling role in advanced manufacturing systems: An emerging human factors domain. *Human Factors, 31*, 635-666.

Sarter, N. B. (1994). *Strong, silent and out-of-the-loop: properties of advanced (cockpit) automation and their impact on human-automation interaction*. Columbus, OH: Unpublished Doctoral Thesis, The Ohio State University.

Sarter, N. B. (1995). *Strong, silent and out-of-the-loop: properties of advanced (cockpit) automation and their impact on human-automation interaction (CSEL Report 95-TR-01)*. Columbus, OH: Cognitive Systems Engineering Laboratory, The Ohio State University.

Sarter, N. B. (1996). From quantity to quality, from individual pilot to multiple agents: Trends in research on cockpit automation. In R. Parasuraman & M. Mouloua (Eds.). *Automation and Human Performance*. Hillsdale, NJ: Erlbaum.

Sarter, N. B. & Woods, D. D. (1991). Situation awareness: A critical but ill-defined phenomenon. *International Journal of Aviation Psychology, 1(1)*, 45-57.

Sarter, N. B. & Woods, D. D. (1992). Pilot interaction with cockpit automation I: operational experiences with the flight management system. *International Journal of Aviation Psychology, 2*, 303-321.

Sarter, N. B. & Woods, D. D. (1994a). Pilot interaction with cockpit automation II: An experimental study of pilot's model and awareness of the flight management system. *International Journal of Aviation Psychology, 4*, 1-28.

Sarter, N. B. & Woods, D. D. (1994b). Decomposing Automation: Autonomy, Authority, Observability and Perceived Animacy. In: M.

Mouloua & Parasuraman, R. (Eds.), *Human performance in automated systems: Current research and trends*. Hillsdale, N. J.: Erlbaum.

Sarter, N. B. & Woods, D. D. (1995). "How in the world did we get into that mode?" Mode error and awareness in supervisory control. *Human Factors*, 37: 5-19.

Sarter, N. B. & Woods, D. D. (1998). Teamplay with a powerful and independent agent: A corpus of operational experiences and automation surprises on the Airbus A-320. *Human Factors*, in press.

Sarter, N. B., Woods, D. D. & Billings, C. E. (1997). Automation surprises. In G Salvendy (Ed.), *Handbook of human factors/ergonomics* (2nd edition). New York: Wiley.

Scardina, J. A., Simpson, T. R., Ball, M. J. (1996, March). ATM: The only constant is change. *Aerospace America*, *3*, 20-23.

Schön, D. A. (1983). *The reflective practitioner: How professionals think in action*. New York, NY: Basic Books.

Schwartz, F. (1998). Joint Strike Fighter: Design and economic issues. *Paper presented at Flygteknik '98*, October, Stockholm, Sweden.

Seamster, T. L., Boehm-Davis, D. A., Holt, R. W. & Schultz, K. (1998). *Developing Advanced Crew Resource Management (ACRM) training: A training manual*. Washington DC: Federal Aviation Administration, Office of the Chief Scientific and Technical Advisor for Human Factors.

Seamster, T. L. & Kaempf, G. L. (in press). Identifying resource management skills for airline pilots. In E. Salas, C. A. Bowers & E. S. Edens (Eds.), *Resource Management in Organisations: A Guide for Training Professionals*. Mahwah, NJ: Lawrence Erlbaum Associates, Inc.

Seamster, T. L., Prentiss, F. A. & Edens, E. S. (1997). Methods for the analysis of CRM skills. *Ninth International Symposium on Aviation Psychology*. Columbus, OH: The Ohio State University.

Seamster, T. L., Redding, R. E. & Kaempf, G. L. (1997). *Applied cognitive task analysis in aviation*. Aldershot, UK: Ashgate Publishing Limited.

Segal, L. D. (1990). Effects of aircraft cockpit design on crew communication. In E. J. Lovesey (Ed.), *Contemporary Ergonomics 1990* (pp. 247-252). London: Taylor & Francis.

Serfaty, D., Entin, E. E. & Deckert, J. C. (1994). Implicit co-ordination in command teams. In A. H. Lewis & I. S. Levis (Eds.), Science of command and control: Part II/Coping with change. Fairfax, VA: AFCEA International press.

Shattuck, L. G. (1995). *Communication of intent in distributed supervisory control systems*. Unpublished Doctoral Dissertation. The Ohio State University.

Sheets, R. G. (1995). *An industry-based occupational approach to defining occupational/skill clusters.* Washington, DC: U.S. Department of Labor.

Sheridan, T. B. (1976). Toward a general model of supervisory control. In T. B. Sheridan & J. Johannesen (Eds.), *Monitoring behavior and supervisory control.* New York, NY: Plenum, pp. 271-282.

Sheridan, T. B. (1987). Supervisory control. In G. Salvendy (Ed.), *Handbook of human factors.* New York, NY: Wiley.

Sheridan, T. B. (1988). Task allocation and supervisory control. In M. Helander, (Ed.), *Handbook of Human-Computer Interaction.* North-Holland, the Netherlands: Elsevier Science Publishers, BV.

Sheridan, T. B. (1992). *Telerobotics, automation, and human supervisory control.* Cambridge, MA: MIT Press.

Sherman, P. J. (1997). *Aircrews' evaluations of flight deck automation training and use: Measuring and ameliorating threats to safety (Technical Report 97-2).* The University of Texas Aerospace Crew Research Project.

Sherman, P. J., Helmreich, R. L. & Merrit, A. C. (1997). National Culture and Flight deck Automation: Results of a Multi-nation Survey. *International Journal of Aviation Psychology, 7.*

Simon, H. A. (1972). *The sciences of the artificial.* Cambridge, MA.: MIT Press.

Singer G. (1998), *FMS pilot evaluation report,* SAAB internal report. Linköping, Sweden.

Smith, P. *et al.* (1996). *Human-centered technologies and procedures for future air traffic management: A preliminary overview of 1996 studies and results (CSEL report submitted to NASA-Ames).* Columbus, OH: Department of Industrial and Systems Engineering, Cognitive Systems Engineering Laboratory.

Smith, K., Scallen, S. F., Knecht, W. & Hancock, P. (1998). An index of dynamic density. *Human Factors, 40(1),* 69-78.

Smith, P. J., McCoy, E., Orasanu, J., Denning, R., Van Horn, A. & Billings, C. E. (1995). *Co-operative problem-solving in the interactions of airline operations control centres with the national aviation system (Chief scientific and technical advisor for human factors report).* Washington, D.C.: Federal Aviation Administration.

Smith, R. E. (1993). *Psychology.* West Publishing Company.

Stassen, H. G. (1986). Decision demands and task requirements in work environments: What can be learned from human operator modelling. In E. Hollnagel, G. Mancini & D. D. Woods (Eds.), (1986). *Intelligent decision support in process environments.* Berlin: Springer Verlag.

Stein, W. (1992). *Models of the human operator in the control of complex systems.* Research Institute for Human Engineering, Wachtberg-Werthoven, Germany.

Steiner, G. (1978). *Heidegger.* Hassocks, Sussex, UK: John Spiers.

Stix, G. (1991, July). Along for the ride? *Scientific American, 7,* 94-105.

Tannenbaum, S. I. & Yukl, G. (1992). Training and development in work organisations. *Annual Review of Psychology*, 43, 399-441.

Taylor, F. W. (1911). *The principles of scientific management.* New York: Harper.

Transport Canada. (1997). *On-board navigation data integrity: A serious problem.* Report of the Transport Canada Database Working Group. Ottawa.

Umbers, I. G. (1979). Models of the process operator. *International Journal of Man-Machine Studies, 11,* 263-284.

Universal (1996), *1MSP*, Flight Management System Pilot Guide.

Van Gent, R. N. H. W. (1994). *Results of the 1994 FAA datalink project conducted at NLR's research flight simulator (National Aerospace Laboratory report).* Amsterdam, the Netherlands: NLR.

Vanderhaegen, F., Crevits, I., Debenard, S. & Millot, P. (1992). *A dynamic task allocation in air traffic control: Effects on controller's behavior.* 6th European Conference on Cognitive Ergonomics (ECCE-6), Balonfüred, Hungary, September 6-11.

Vicente, K. J., Burns, C. M., Mumaw, R. J. & Roth, E. M. (1996). How do operators monitor a nuclear power plant? A field study. *Proceedings of the 1996 American Nuclear Society International Topical Meeting on Nuclear Plant Instrumentation. Control and Human-machine Interface Technologies,* pp. 1127-1134. LaGrange Park, IL: ANS.

Voss, J. F. & Wiley, J. (1995). Acquiring intellectual skills. *Annual Review of Psychology*, 46, 155-181.

Wason, P. C. & Johnson-Laird, P. N. (1972). *Psychology of reasoning.* London: B. T. Batsford.

Weizenbaum, J. (1976). Computer power and human reason. From judgment to calculation. San Francisco: W. H. Freeman.

Wenkebach, U., Pollwein, B. & Finsterer, U. (1992). Design of a Human Interface for the Analysis of Large Data Sets in Intensive Care. In Lun, K. C. *et. al.* (eds), *MEDINFO '92,* pp. 1267-1272.

Wiener, E. L. (1987). Fallible humans and vulnerable systems: Lessons learned from aviation. In J. A. Wise & A. Debons (Eds.), *Information systems: Failure analysis (NATO ASI Series Vol. F32).* Berlin: Springer-Verlag.

Wiener, E. L. (1988). Cockpit automation. In: E. L. Wiener & D. C. Nagel (Eds.), *Human factors in aviation.* San Diego, CA: Academic Press.

Wiener, E. L. (1989). *Human factors of advanced technology ("glass cockpit") transport aircraft (NASA contractor report No. 177528).* Moffett Field, CA: NASA-Ames Research Center.

Wiener, E. L. (1993). Crew co-ordination and training in the advanced-technology cockpit. In E. L. Wiener, B. G Kanki & R. L. Helmreich, (Eds.). *Cockpit resource management.* San Diego, CA: Academic Press.

Wiener, E. L., Chidester, T. R., Kanki, B. G., Palmer, E. A., Curry, R. E. & Gregorich, S. A. (1991). *The impact of cockpit automation on crew co-ordination and communication: I. Overview, LOFT evaluations, error severity, and questionnaire data* (NASA Contractor Report No. 177587). Moffett Field, CA: NASA-Ames Research Center.

Wiener, E. L. & Curry, R. E. (1980). Flight deck automation: Promises and problems. *Ergonomics, 23* (10), 995-1011.

Wiener, E. L., Kanki, B. G. & Helmreich, R. L. (Eds.) (1993). *Cockpit resource management.* San Diego, CA: Academic Press.

Wiener, N. (1954). *The human use of human beings.* New York: Houghton-Mifflin.

Winograd, T. & Flores, F. (1986). *Understanding Computers and Cognition: A New Foundation for Design,* Norwood, NJ: Ablex.

Woods, D. D. (1992). *Cognitive activities and aiding strategies in dynamic fault management (Series of three Cognitive Systems Engineering Laboratory Reports).* Columbus, OH: Department of Industrial and Systems Engineering, Cognitive Systems Engineering Laboratory.

Woods, D. D. (1993). Process-tracing methods for the study of cognition outside of the experimental psychology laboratory. In G. Klein, J. Orasanu, R. Calderwood & C. E. Zsambok (Eds.). *Decision making in action: Models and methods.* (pp. 228-250). Norwood, N.J.: Ablex.

Woods, D. D. (1994a) Automation: Apparent simplicity, real complexity. *Proceedings of the First Automation Technology and Human Performance Conference, Washington, D.C.,* April 7-8.

Woods, D. D. (1994b). *Visualising Function: The Theory and Practice of Representation Design in the Computer Medium.* Unpublished manuscript. Columbus, OH: Ohio State University.

Woods, D. D. (1996). Decomposing automation: Apparent simplicity, real complexity. In R. Parasuraman & M. Mouloula, (Eds.). *Automation Technology and Human Performance.* Hillsdale, N.J.: Erlbaum.

Woods, D. D. (1997). Avoiding Automation Surprises. *Presentation at SAS Flight Academy,* Stockholm.

Woods, D. D. & Hollnagel, E. (1987). Mapping cognitive demands in complex problem-solving worlds. *International Journal of Man-Machine Studies, 26,* 257-275.

Woods, D. D., Johannesen, L. J., Cook, R. I. & Sarter, N. B. (1994). *Behind human error: Cognitive systems, computers, and hindsight.* (CSERIAC State of the Art Report) Dayton, OH: Crew Systems Ergonomic Information and Analysis Center.

Woods, D. D. & Roth E. M. (1988). Cognitive systems engineering. In M. Helander (Ed.) *Handbook of Human - Computer Interaction*. North-Holland, the Netherlands: Elsevier Science Publishers, BV.

Woods, D. D. & Sarter, N. B. (1993). Evaluating the impact of new technology on human-machine cooperation. In: J. A. Wise, V. D. Hopkin & P. Stager (Eds.), *Verification and validation of complex systems: Human factors issues*. Bonn: Springer-Verlag.

Woods, D. D. & Sarter, N. B. (1998). *Learning from automation surprises and "going sour" accidents: Progress on human-centered automation (report ERGO-CSEL-98-02)*. Columbus, OH: Institute for ergonomics, The Ohio State University.

Index

A

C

D

E

W